Precious Metals, Coinage and the Changes of Monetary Structures in Latin-America, Europe and Asia

STUDIES IN SOCIAL AND ECONOMIC HISTORY
Edited by Herman Van der Wee

Volume II

Eddy H.G. Van Cauwenberghe ed.

Precious Metals, Coinage and the Changes of Monetary Structures in Latin-America, Europe and Asia

(Late Middle Ages - Early Modern Times)

Proceedings of the "Stocton Conference on Monetary History"
organized in October 1985
at the University of the Pacific, Stocton / CA.
by Professor Dennis O. Flynn

Leuven University Press
1989

C.I.P. KONINKLIJKE BIBLIOTHEEK ALBERT I

Precious

Precious metals, coinage and the changes of monetary structures in Latin-America, Europe and Asia : late middle ages - early modern times / editor Eddy H.G. Van Cauwenberghe.
Leuven : University Press, 1989. – 222 p. : ill.; 24 cm.
(Studies in social and economic history; vol. 2).
ISBN 90-6186-335-X
SISO : 902.9 UDC : 737"13/18"

Onderwerpen : Numismatiek

D / 1989 / 1869 / 34

Printed in Belgium

CONTENTS

PREFACE

On August 17, 1982 the Session C–7 «Minting, Monetary Circulation and Exchange Rates», was organized in Budapest at the «8th International Economic History Congress». A large group of specialists gathered in Budapest to report on current research and to discuss provisional and final results that had been achieved in the field of monetary history. The papers given, revised and expanded, were the basis for a book published in 1984 by the THF (Verlag Trierer Historische Forschungen, Trier/W. Germany). In addition, this book includes the papers presented at the Bloomington - S.S.H.A. Conference, held in November 1982.

At the time no one could predict that this was the start of a solid tradition which would bring us together three years later at the University of the Pacific at Stockton/CA and last year (June 1987) at Keio University in Tokyo, Japan.

Most of the contributions to these first conferences deal with money and exchange from as early as the twelfth century until the nineteenth century, and the focus of attention is northwestern Europe. However, a couple of essays on the monetary history of Canada, the United States and India are also included. The specific issues under discussion are the size and form of specie production and circulation, the history of mint standards and procedures, relationships among money systems and mints, the function of money in the age before official statistics, and the quantity theory of money.

The third conference was hosted by Professor Dennis O. Flynn in October 1985 in Stockton. Together we selected a distinguished international group of scholars of varied experience and reputation and in distinctive disciplinary specialities. The resulting publication of the Stockton proceedings contains several major contributions to the monetary history of Europe, America and Asia.

In his short but remarkable paper, Professor Michel Morineau offers «food for thought» to anyone dealing with the history of precious metals, money and capital. He stresses the necessity of «pooling» knowledge, cooperation and the constructive exchange of information. Therefore let us hope that the continuation of these conferences, together with their growing importance, may contribute to the progress of international collaboration in the field of monetary history.

The Stockton papers concerning ***Europe*** deal with the problem of ***monetary production and the possibilities of the reconstruction of the coin (money) stock*** as presented in the contributions from Professor Eddy Van Cauwenberghe and Dr. Rainer Metz and from Dr. Joachim Schüttenhelm. On the other hand, European monetary systems and the economies of medieval Europe are discussed in a substantial study by Professor John H. Munro on the ***Petty Coinage*** (fractional coinage) in the economy of late-medieval Flanders.

Other contributors discuss the ***problem of transfer of precious metals and specie flows***. ***Bullion transfer from Western Europe to the Baltic and the problem of trade balances (1550–1750)*** is studied by Dr. Michael North and by Professor Carlo M. Cipolla in the latter's contribution ***American Treasure and the Florentine Coinage in the Sixteenth century***. Dealing with the same question of international bullion flow, the ***Bullion flow from Europe to the East (1500–1800)***, Professor Arthur Attman explains how very large parts of the imports of American precious metals into the Iberian peninsula were transferred and dispersed to a great extent to the Baltic, the Levant and Asia. Precious metals also came to Asia from America via the Pacific Ocean and Manila. From such information, it becomes more evident that rather than focus on money supplies and individual countries, the analysis concerns world demand for and world supply of monetized and non-monetized precious metals combined, as Professors Dennis O. Flynn and Kerry W. Doherty explain in their theoretical reflections on the ***Microeconomic Quantity Theory of Money and the Price Revolution***.

Investigations done by Professor Alan K. Craig on ***Mining Ordenanzas and Silver Production at Potosi (1545–1578)*** in the municipal archives of Arequipa, offer new empirical evidence on the question of mine production estimates and thus on the evolution of cost of production rates.

It is clear from the literature that existing estimates of the world stock of metallic money cannot be more than rough estimates for the prestatistical period; we believe that the subject of quantitative monetary history can only be approached by combining theoretical and empirical evidence. The importance of theory and methodology when investigating historical phenomena has clearly been demonstrated by Professor Winfried Stier in his paper ***Meaning and Functions of New Methods of Time Series Analysis for Economic History***. The advantages of the use of filter-design, based on modern filter-theory, for the analysis of historical time series and for the detection of any possible long cycles are obvious.

Contributions to ***Asian monetary history*** in the Stockton conference volume are made by Dr. Frank Perlin and by Professor Kazui Tashiro. Perlin's study on the ***Division of Labor in European and Indian coin manufacture before mechanisation***, not only shows the interesting possibility of a comparison between European and Asian evidence, but especially demonstrates the many advantages offered by an analysis of the contrasts and the resemblances between two regions (Western Europe and India). The interest of a comparative approach becomes even more evident from the communication by Professor Dr. Kazui Tashiro on ***Exports of Japan's Silver to China via Korea and Changes in the Tokugawa Monetary System during the 17th and 18th Centuries***. The author focusses on the flow of silver in the

Japanese-Korean trade, and shows quantitatively how this flow affected commerce between Korea and China and the monetary system in Japan.

It is clear that the success of the Stockton conference was dependent first of all on the scientific programme and therefore we want to thank all of the participating individuals, authors and those who contributed to the discussions during the conference. Our appreciation also goes to those who provided the necessary technical assistance and material support: the local Stockton Organising Committee at the Department of Economics, and the people who organised the social events for this occasion. In the name of all the participants, I have the pleasure to thank especially our friends Dennis and Nancy Flynn for their enthusiasm and for their energy and support in making this undertaking possible. Also we want to express our sincere gratitude to all our collaborators at UFSAL/Brussels (Universitaire Faculteiten Sint Aloysius) who were involved in the preparations and the publication-process: Drs. Wenceslaus Mertens, Dr. Erika Meel, Drs. Myriam Carlier and Ms. Carla Morlion. Finally, we are grateful to Dr. Rainer Metz for his help in transforming the manuscript into a book.

We hope that for all concerned these proceedings may have proved to be a challenging and stimulating exposure to new materials and further collaboration.

Eddy H. G. Van Cauwenberghe

COINAGE AND THE COIN (MONEY) STOCK: PROBLEMS, POSSIBILITIES AND FIRST RESULTS (THE SOUTHERN LOW COUNTRIES 1334-1789)

by Eddy van CAUWENBERGHE and Rainer METZ

The main purpose of this short paper is the attempt to open up a discussion by presenting a first estimation of the money stock in the Southern Low Countries on the basis of the mint production figures. The empirical basis for this estimation are the mint accounts of the mints in Antwerp, Bruges, Brussels and Gent. An empirical investigation of this problem is confronted with mainly two serious questions:

1. How reliable is the existing data material?
2. How can one get valid empirical numbers for the theoretical key variables, like money stock, general price index etc.?

Despite the large number of investigations dealing with this matter, in our opinion a new effort is worthwile primarily because of the new and remarkable mint production statistics. With this material we have for the first time the possibility to calculate a lot of monetary variables which are of great importance to the analysis of monetary structures and developments.

We have already described the structure of the data and the way they are evaluated in an article published in 1984[1], so that we can confine ourselves to a brief summary.

The material consists of mint master accounts dating back to the middle of the 14th century. Thus the source we use is the official bookkeeping of the masters of the different mints having operated in the Low Countries between 1334 and 1789. The accounts of the mint masters were subjected to the severe control of the «Chamber of Accounts», a central institution introduced in 1386 by the young Burgundian administration.

At the moment the edition of the whole material is in preparation[2]. This publication will contain a lot of variables in the form of tables and graphs. So for example:

[1] E. van CAUWENBERGHE, R. METZ: *Geld und Währung in den südlichen Niederlanden während der Frühen Neuzeit*. In: Trierer Historische Forschungen, Vol. 7, Trier 1984, p. 123–150.

[2] E. van CAUWENBERGHE, F. IRSIGLER, R. METZ, J. H. MUNRO: *Coinage in the Low Countries. 14th–18th Centuries*, Vol. 1 – Antwerp, Bruges, Brussel, Gent; forthcoming.

1. The fine- and rough weight, taille, rate etc. for each coin occurring in the accounts.
2. The minted fine weight, the total nominal money, the total number of coins minted etc. as aggregated figures for each minting period and for each coin.
3. Additionally some of these variables have been aggregated for all types of coins both per minting periods as well as yearly figures, for example total weight of minted silver and gold (fine and rough weight), the total amount of nominal money in pound groat, etc. Because of the fact that these figures provide rather macro-economic variables, they serve as basic material for the following considerations.

These considerations are divided into three parts:

(1) Representing and describing the secular movement.
(2) Modelling and estimating the coin stock.
(3) Discussing preliminary results.

The secular movement

Before entering into the discussion of the theoretical model, we might have a look to the secular movement of some of the aggregated monetary variables derived form the statistics of mint production. These long term movements will serve as empirical framework for the following analysis and provides a good impression of general tendencies. Because of the dominance of the Antwerp mint, we confine ourselves to the Antwerp mint statistics.

In a long term view the coin production in Antwerp, which is indicated by the trend of the nominal value of the production in terms of money of account (Fig. 1)[3], shows a first considerable increase at the end of the 16th century, then, after the middle of the 17th century, it becomes insignificant. This first increase had been preceded since 1540 by a phase of expansion, which lasted about 30 years and broke off in 1570.

If the amount of money of account is compared with the development of the quantity of minted precious metal, the relationship between the two series reveals an inflationary development (Fig. 2)[4], which assumes huge

[3] The trend is estimated with a filter method developed by R. Schmidt and W. Stier, which allows to define the trend as specific frequencies. In our case all frequencies with a duration from infinite to 25 years are defined as trend components. For the filter method see R. Schmidt: *Konstruktion von Digitalfiltern und ihre Verwendung bei der Analyse ökonomischer Zeitreihen*. Bochum 1984; R. Metz: *Ansätze, Begriffe und Verfahren der Analyse ökonomischer Zeitreihen*. In: Historical Social Research/Historische Sozialforschung, Vol. 13, 1988, No. 3, pp. 23–103.

[4] With the ratio of the amount of nominal money to the weight of precious metal – here called «inflation index» – one can demonstrate in what way an increasing

proportions in 1550 and does not stabilize until 1615. During this period the inflation index rose from about 8.4 to 16.4 (95%)[5].

Nearly contrary to these tendencies is the relation between the average fine weight and rough weight of minted coins[6]. This quotient declines constantly until 1520 with regard to silver coins (Fig. 3) and until about 1540 as far as gold is concerned (Fig. 4). What is clearly expressed here, is the continuous scarcity of precious metal, by which the Late Middle Ages are characterized. Only the influx of overseas precious metal brings about a turn, which manifests itself in a rising ratio. From 1550 already the minting relation for silver is characterized by a high value, which apart from 1590 to 1630 remains nearly constant for the rest of the period under investigation. The ratio for gold also changes almost abruptly from 1540 to settle down on a nearly constant level at the beginning of the 17th century. The relatively erratic transition periods correspond, with regard to both gold and silver to the phases of the enormous depreciation of the money of account. Obviously this inflation was not caused by a debased coinage but by a successive increase of the rates (nominal value) of very fine minted coins[7].

A comparison between the trend of the deflated price index[8] and the trend of the quantity of minted precious metal (Fig. 5 is showing both curves)

---

amount of nominal money is produced from a constant wheight quantity of the precious metal. Figure 2 shows the long term development of this relation for the important mint of Antwerp between 1474 and 1712. Moreover it can be shown that this relation which is nothing else than an averaged index for all coins, corresponds to the reciprocal value of the fine weight equivalent of the money of account.

[5] In passing we would like to mention that this increase corresponds to the example which F. C. SPOONER: *International Economy and Monetary Movements in France 1493–1725*, Cambridge 1972, p. 92, gave for France with regard to the Malestroict-Bodin controversy, since in Spooner's example the French price index for the period of 1555 to 1590–98 also rose by approximately 100% (from 265.2 to 627.5 index points).

[6] This relation was obtained by dividing the total quantity of minted fine metal through the total quantity of minted rough metal.

[7] A paragon in this context is the mint policy under Phillip II. For example the «Philippusdaalder» minted in Antwerp was 1555 issued with a face value of 66 groate. Without modifications in taille or fine weight this face value was raised to 100 groate in the year 1585. The consequence of such a policy was the depreciation of the fine weight equivalent of the money of account. According to the raising face value of the «Philippusdaalder», and the other coins respectively, the fine weight of the groat – as money of account – diminished from 0.432 to 0.286 grams and this without a reduction of the fine weight of the coins.

[8] For the price index see H. VAN DER WEE: *Prijzen en lonen als ontwikkelings variabelen. Een verglijkend onderzoek tussen Engeland en de zuidelijke Nederlanden 1400–1700.* In: Album Charles Verlinden. Gent 1975, p.413–447. The price index was deflated by the fine weight equivalent of the money of account, calculated from the Antwerp mint series; compare also E. VAN CAUWENBERGHE, R. METZ: *Geld und Währung in den südlichen Niederlanden während der Frühen Neuzeit.* In: Trierer Historische Forschungen, Vol. 7, Trier 1984, p. 137.

shows that from about 1500 there is a rise in the price index parallel to the increase in coinage output, while the development until 1500 is characterized by a downward trend. Over the whole period from 1500 to 1630/40 both curves express a long term growth, which from 1640 is followed by a depression, again the courses of both curves being parallel.

A similar development with regard to the 16th and 17th centuries also emerges from a comparison between the trend of nominal prices[9] and the trend of the amount of money of account. It is remarkable that the strong inflation, or in other words the strong depreciation of the money of account, is accompanied from about 1570 onward by a rapid increase in price inflation. This connection becomes evident when the rising price level is compared with the inflation index for silver coins (Fig. 6).

Problems and Possibilities of Estimating the Coin (Money) Stock

The Relationship Between Mint Output and Coin (Money) Stock

None of the existing concepts of quantifying the volume of money in prestatistical periods allows a reliable estimation of the money stock.

For the Southern Netherlands we have, in contrast to all other countries, an exceptional and nearly unbroken series of mint accounts or mint delivery books from all mints in the region between 1334 and 1789. In addition we have some evidence from coin hoards, 250 hoards distributed over a period of 550 years. However, these hoards are for the most part incompletely inventoried; sometimes even the number of coins is unknown.

Therefore the most useful method seems to be the concept of mint-production[10], i.e. the money stock is defined as the volume of money in coined form, thus as the domestically minted coins in circulation in the Southern Low Countries. We know this ignores the obvious complications caused by, e.g. bills of exchange, and, probably more seriously, the legal circulation of foreign coins, because mint records provide no information about coin import or export; finally there is the problem of losses.

---

[9] See n. 8.

[10] A first discussion of this model in E. VAN CAUWENBERGHE; D. HAENECAERT: *Mintage and Coin Circulation in the Southern Low Countries (14th – 18th Centuries) – Some Theoretical Reflections –*. In: Trierer Historische Forschungen, Vol. 7, Trier 1984, pp. 151–180.

Because the problems of loss (constant and variable) remain unsolved by the mint-production concept itself, loss rates must be estimated by other methods. In both cases, variable - and constant loss, the issue is the loss of a part of the coin stock. By variable loss we mean losses fluctuating in amount. By constant loss we mean losses at a steady or continuous rate.

As causes of variable losses authors usually refer to capital exports and with less emphasis, at least for the Southern Low Countries, shipwrecks and the burial or simply the loss of coins. In a detailed study of this problem Mayhew writes: «It is not possible to provide even approximate figures for such losses»[11].

With constant losses the problem arises from the material form of the money. A coin derived its extrinsic value from its contents, namely the quantity of fine silver or gold that it contained. This means that theoretically any loss in weight in the coin is transposed into a loss in value: the rate falls. Such losses had two principal causes: the natural wear and tear from the use of the coin and fraudulent actions.

Constant loss is, however, a dynamic process. The different factors that influence this process are interrelated; they are synchronic and diachronic in their effect. Moreover, constant loss is a cumulative process: the quantity of wear and tear during one period largely determines the wear and tear of each subsequent period.

It is therefore impossible to speak of constant losses in general, a specific form of constant loss exists for each mint type, in fact for each coin, each period, and each region. No single method of general applicability can be developed for measuring the loss rate, not even for a specific period or coin type. Thus, it is apparent that the estimates of loss due to wear and tear vary in the literature.

Methods of coin find- ananlysis may provide indications on a global loss rate (variable and constant loss). But, there are many risks associated with the use of existing techniques on the coin hoards from the Southern Netherlands.

If, however, the aim is to arrive at an estimate of the coin stock, one must take this phenomenon into account in one way or another: a more general form of the loss rate is required.

Among others C.C. Patterson has published a mathematical model for the calculation of the constant loss rate[12]. Elsewhere we also investigated a more general form for solving this problem[13]. On the basis of different hypotheses concerning the process of constant loss, we drafted several mathematical functions. Our conclusion was that the constant loss rate indeed differs for

---

[11] N. J. MAYHEW: *Numismatic Evidence and Falling Prices in the Fourteenth Century.* In: The Economic History Review, 2nd series, 27(1)/1974, p. 4.

[12] C. C. PATTERSON: *Silver Stocks and Losses in Ancient and Medieval Times.* In: The Economic History Review, 2nd series, 25(2)/1972, pp. 205-233.

[13] See n. 10.

each coin type but that, if the period is sufficient long, a single constant rate may be assumed as a general form. The counterbalancing effects of different wear and tear patterns over a long period can always give rise to a single constant rate. Moreover, the choice of loss rate clearly affects the levels of the coin stock at any point in the time, but, the rates of growth of the series are clearly much less sensitive to this choise. For the study of the Price Revolution, e.g., rates of change of the money stock are rather important than absolut levels, so that the choice of a loss rate is not as critical as it might appear.

Modelling and Estimating the Coin Stock in the Low Countries

As we already explained, direct mint output data for a period of more than 400 years will be used for the estimation. We may accept that they can provide for the first time a systematic idea of the evolution of the coin stock in the Low Countries. How accurate that idea will be depends mainly, despite the imperfections characterizing each possible concept, on the techniques used and thus is a question of method.

Since we already published our method for estimating the coin circulation from a mathematical model, we only repeat here the most essential steps.

The challenge was to built a model that would permit, instead of a static estimation, a systematic and dynamic evaluation of the coin stock, based on the absolute data of the coinage. This approach differs from the attempts made before by historians: previous writers have used mint output data in constructing money stock estimates, implicitly assuming that annual mint output is a proxy for annual changes in the money stock. F. Spooner, e.g., used thirty and sixty year moving averages of mint output to generate a money stock series for France[14].

We used a highly simplified model, a simple mathematic equation based on a couple of logical assumptions:

1. the global coin circulation, at a given moment, is not exclusively but largely dependent on monetary production from previous periods;
2. part of the production from each previous period disappears, and there exists a strong relationship between the volume of these losses and the number of years the given production served as currency.

---

[14] F. C. SPOONER: *International Economy and Monetary Movements in France 1493–1725*, Cambridge 1972.

The mathematical form of the model can be expressed by this equation:

$$C_t = (1-a)^1 \; P_{t-1} + (1-a)^2 \; P_{t-2} + (1-a)^3 \; P_{t-3} + \ldots (1-a)^i \; P_{t-i}$$

With:
Ct = the estimated coin stock at the beginning of period t
a = the estimated annual rate off loss or the «coefficient of deterioration»
P_{t-i} = the production of period t-i
$(1-a)^i$ = the «Theoretical Final Value» (T.F.V.)

thus:

$$a = 1-(\text{TFV})^{1/i}$$

In this equation, P represents a given variable; both «a» and «i» are parameters. Values for P (coinage) are available for more than 400 years. The remaining problem is to give values to the parameters.

For this purpose we insert the notion Theoretical Final Value. In principle the T.F.V. represents the final weight of a coin - at the moment it goes out of circulation - compared to its initial weight. The T.F.V. we determined takes into account every kind of loss in the global production and even the entire disappearance of coins from circulation in a given period. We already developed and published our arguments for postulating, with respect to the Southern Low Countries, a general T.F.V. of 0.35 for a period of 30 years. We know that this remains clearly an assumption. Given the form of our model, however, the postulated value has only a smaller effect on the ultimate results obtained. Even, if for the estimated annual values adjustments always may be necessary, the fundamental trend is still indicated.

On the basis of this T.F.V. a value for a may be derived.

Example:

TFV = 0.35 for i = 30 years, i.e.
$0.35 = (1-a)^{30}$
$a = 1 - (0.35)^{1/30}$
$a = 0.0344\ldots = 3.44\%$

$$C_t = (1-0.0344)^1 \; P_{t-1} + (1-0.0344)^2 \; P_{t-2} \, (1-0.0344)^3 \; P_{t-3} \ldots + (1-0.0344)^{30} \; P_{t-30}$$

It is clear that the Real Final Value underwent various changes through the time. Also it is true that it was different for gold and silver, and for each specific type of coin.

The principal cause of this variation was the different regard people had for coins of different value. The apparent relationship between coin size and loss is superficial. In addition to the general T.F.V. there is to ascertain a specific value for gold and silver. By fixing two Theoretical Final Values for each minted metal, two curves are obtained for each metal. The effective stock of coins will be found between each set of curves. For each period the results from this procedure can be refined on the basis of qualitative an quantitative data or by complementary information obtained from the analysis of coin hoards.

The Application of the Model: First Results

Figures 7, 8, 9 and 10 show the estimated coin circulation, based on the proposed model. Figures 7 and 8 are plotted on semi-logarithmic scale, both figures 9 and 10 represent absolute values;

It must be taken into account that the applied method does not but estimate the circulation of domestic coin-types: the importance of the circulation of foreign coins in the 18th century is surely not to be underestimated. In the first half of the century, they constitute the main part of the total circulation, at the beginning of the second half of the century almost the complete coinage is based on the remelting of foreign coins;

Figure 8 was plotted only to show the effect of gaps in the series; the problem of absence of data for certain years or periods was solved here by giving a 0 value.

The T.F.V. used to compute the series was 0.35 for a period of 30 years and this for both silver and gold coins.

- It is very clear (Figure 7 and 8) from the production and from the circulation figures that gold and silver output alternate during most of the observed period. Only at the beginning and at the end there seems to be a strong parallelism between both curves; also there might exist a cyclical pattern in the course of both production and -circulation (Figures 7, 9 and 10).

From ca. 1470 until ca. 1510 silver made a dominant contribution to the circulation:

- silver flowed during the second half of the 15th and early 16th centuries into the Low Countries from Germany, Hungary and Tirol;
- the gold-silver ratio evolved from 1:9 to 1:12 in the course of the 16th century;

From ca. 1510 until ca. 1565 gold dominates:

- it is the strongest episode of the «Price Revolution» in the Low Countries and the main period of commercial expansion (Antwerp);

- there is a net increase of the importation of precious metals from the New World;
- the gold-silver ratio is rising gradually to about 1:15, but governmental monetary policy favored the circulation of gold coins;

From ca. 1565 until ca. 1600 again we observe a dominant silver circulation:
- the Low Countries' «Price Revolution» is abating;
- the huge influx of precious metals, mainly silver, provokes a remarkable increase of silver minting during the third quarter of the 16th century, now more and more on the basis of imports of American silver from Spain: massive imports into the Southern Low Countries are done by the Spanish government in order to finance war (Revolt);

From ca. 1599 until ca. 1612, gold again dominates:
- special measures are taken by the government of the archdukes, Albrecht and Isabella, reforming the whole currency system and favoring gold at the moment that the gold price was rising further.

From ca. 1612 until ca. 1712 we note a dominance of silver circulation:
- the general currency reform of 1612/1618 was favoring silver relative to gold at the moment that the relative silver price at the market was declining;
- the creation of new heavy silver coins stabilized the monetary circulation;
- the statistics (Figure 10) of the Antwerp mint, give a clear illustration of the impressive expansion of the silver circulation caused by the dominant heavy silver coins;

From 1712 there is a parallelism in the lowering trend of both gold and silver circulation:
- between 1710 and 1749 there was hardly any production;
- from 1749 production resumed but with almost immediate decreasing trend;
- coinage was fed by the remelting of foreign silver and gold coins;
- there was a negative balance of payments, the stock of foreign money was probably exhausted at the beginning of the 1770's.

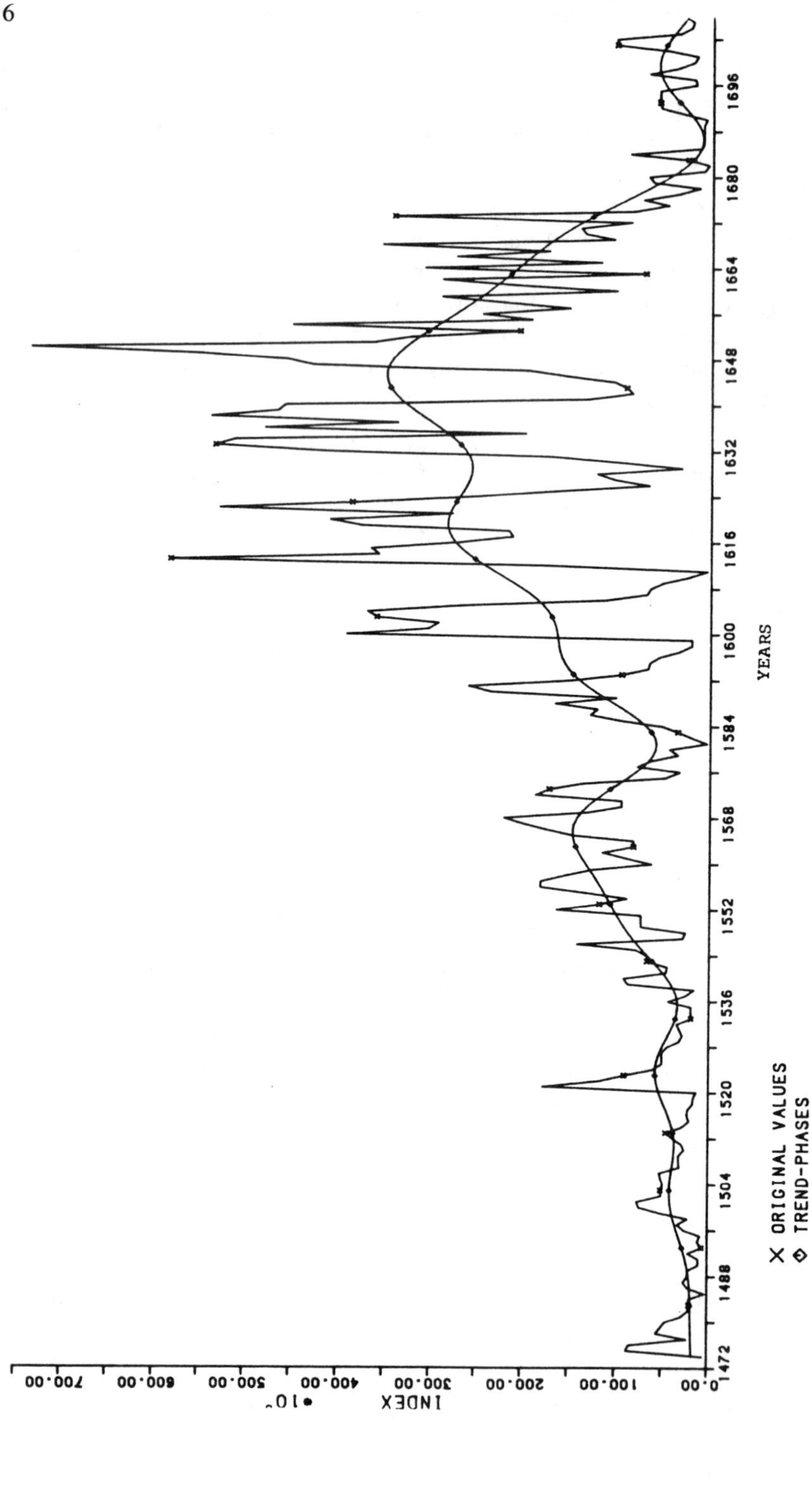
INDEX *10^
0.00
100.00
200.00
300.00
400.00
500.00
600.00
700.00
1472
1488
1504
1520
1536
1552
1568
1584
1600
1616
1632
1648
1664
1680
1696
YEARS
X ORIGINAL VALUES
◇ TREND-PHASES

Fig. 1

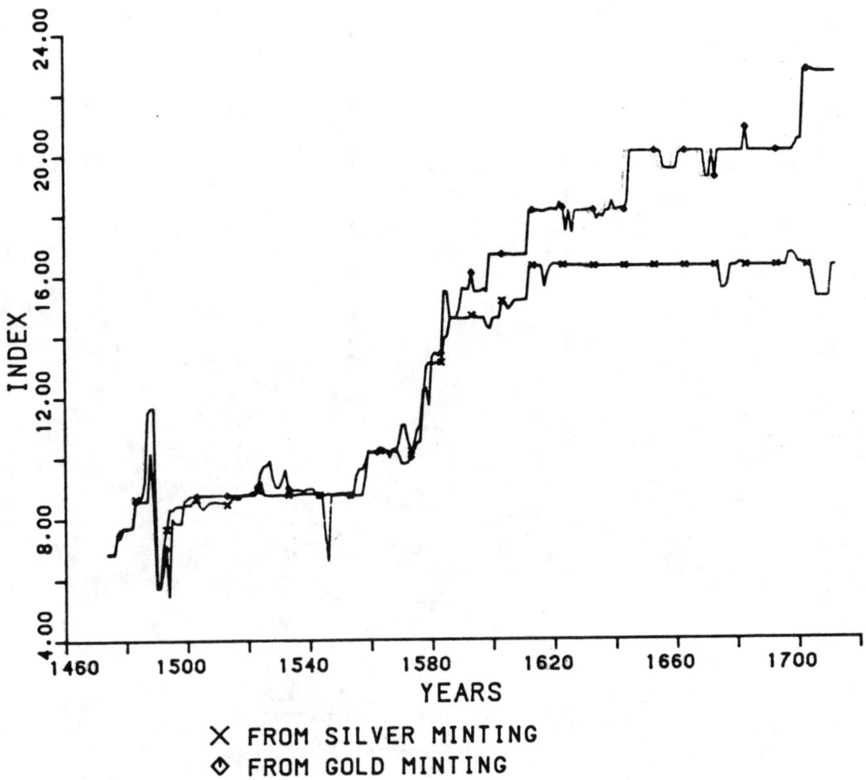
24.00
20.00
16.00
12.00
8.00
4.00
INDEX
1460
1500
1540
1580
1620
1660
1700
YEARS
X FROM SILVER MINTING
◇ FROM GOLD MINTING

Fig. 2

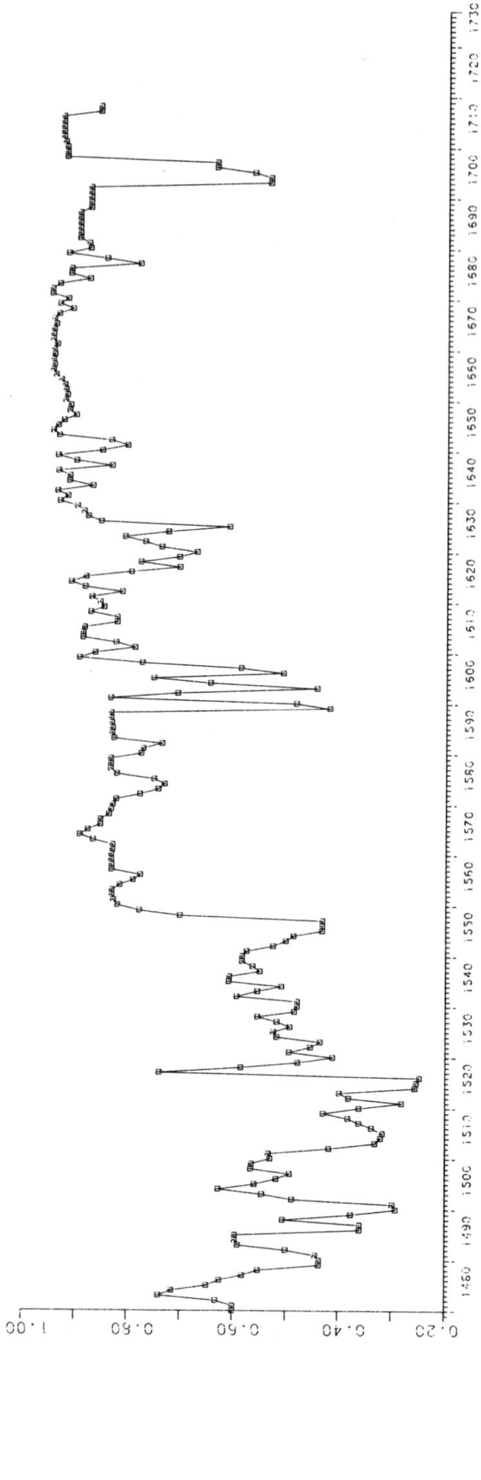
0.20
0.40
0.60
0.80
1.00
1480
1490
1500
1510
1520
1530
1540
1550
1560
1570
1580
1590
1600
1610
1620
1630
1640
1650
1660
1670
1680
1690
1700
1710
1720
1730

Fig. 3

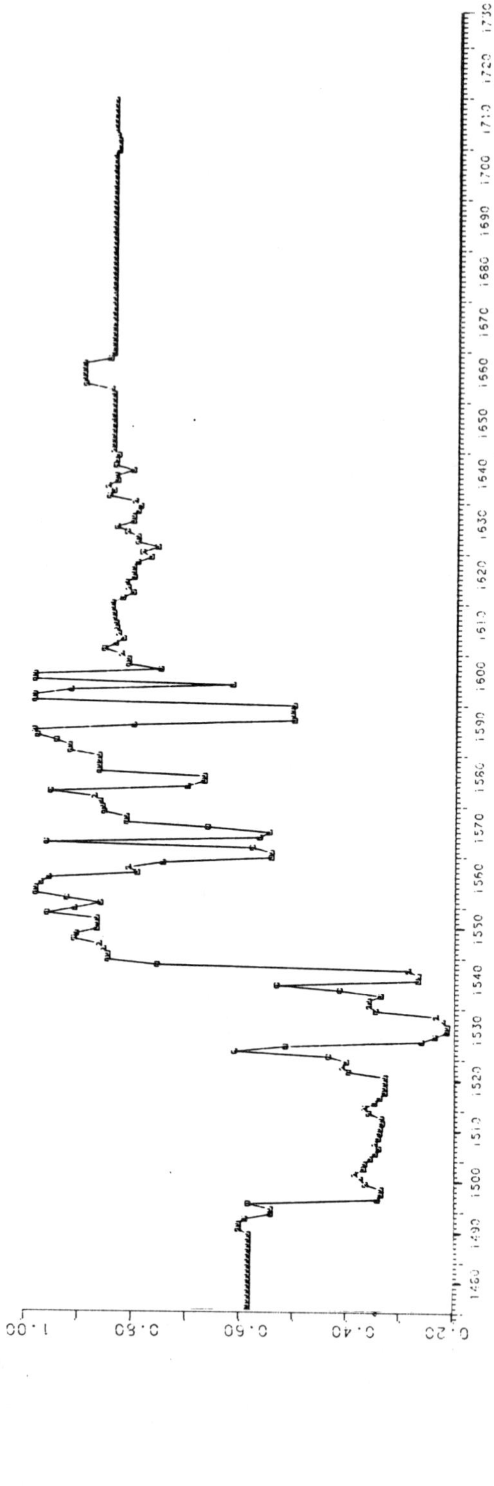
0.20
0.40
0.60
0.80
1.00
1480
1490
1500
1510
1520
1530
1540
1550
1560
1570
1580
1590
1600
1610
1620
1630
1640
1650
1660
1670
1680
1690
1700
1710
1720
1730

Fig. 4

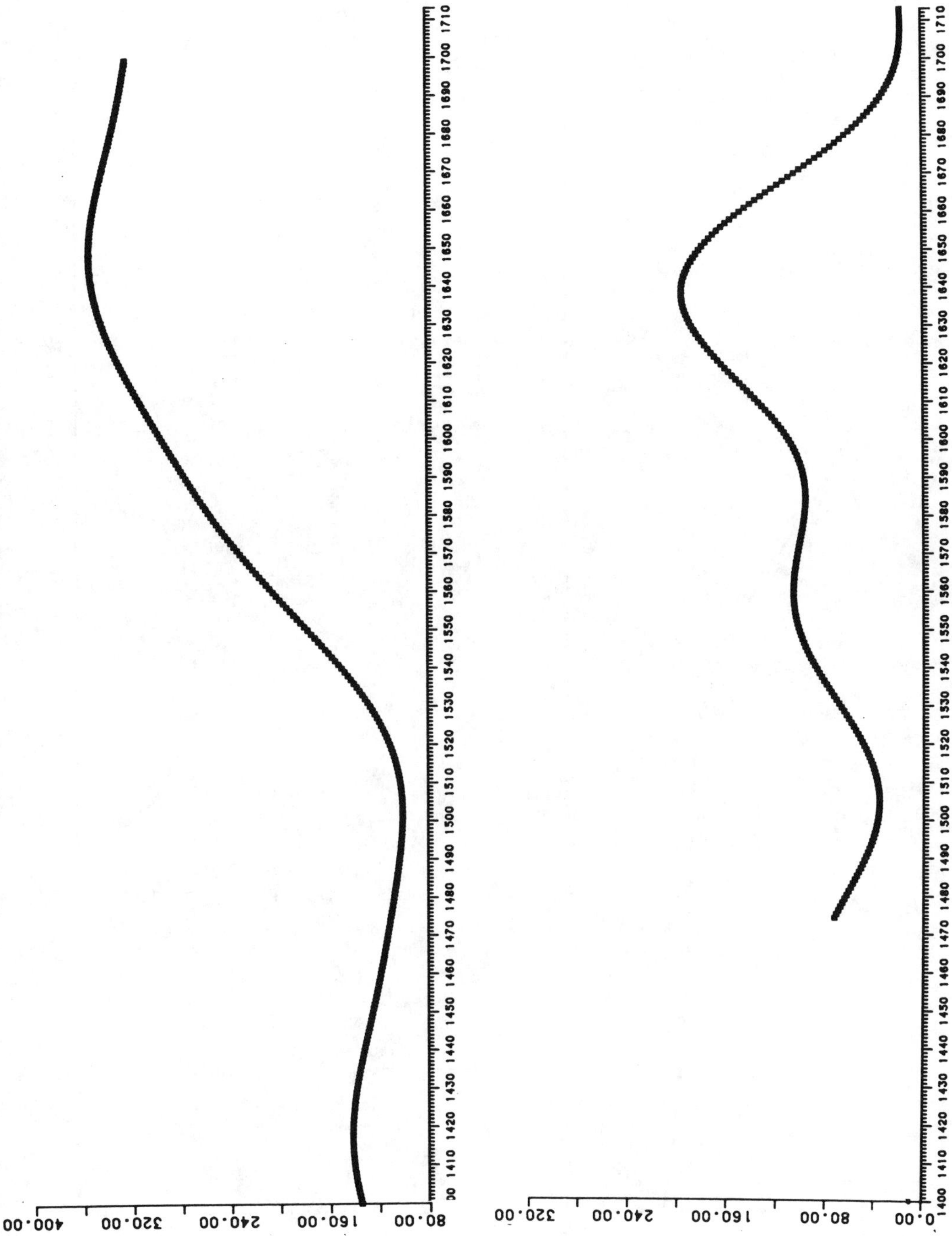
1400 1410 1420 1430 1440 1450 1460 1470 1480 1490 1500 1510 1520 1530 1540 1550 1560 1570 1580 1590 1600 1610 1620 1630 1640 1650 1660 1670 1680 1690 1700 1710
80.00 160.00 240.00 320.00 400.00
0.00 80.00 160.00 240.00 320.00

Fig. 5

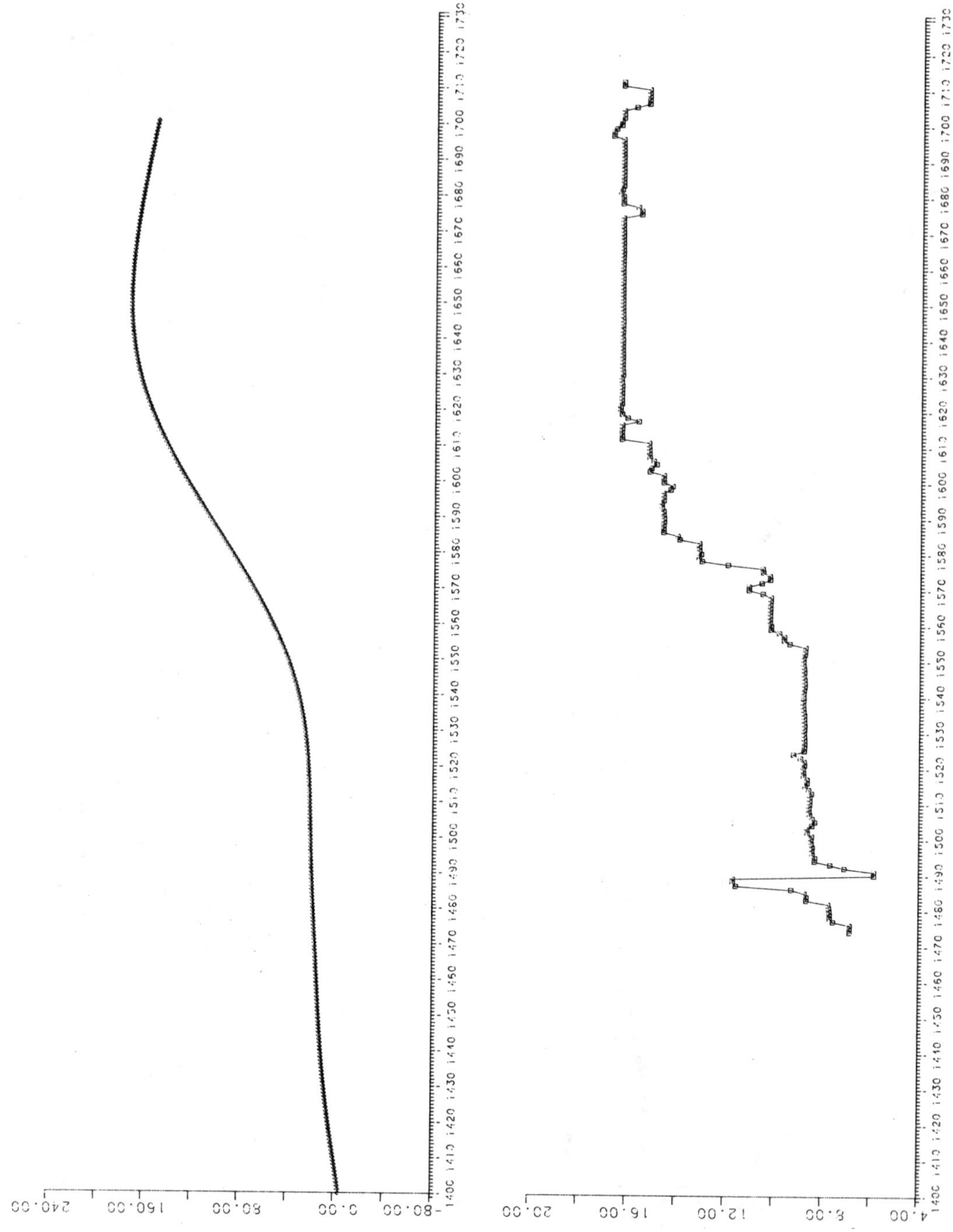
-80.00
0.00
80.00
160.00
240.00
1400 1410 1420 1430 1440 1450 1460 1470 1480 1490 1500 1510 1520 1530 1540 1550 1560 1570 1580 1590 1600 1610 1620 1630 1640 1650 1660 1670 1680 1690 1700 1710 1720 1730
4.00
8.00
12.00
16.00
20.00
1400 1410 1420 1430 1440 1450 1460 1470 1480 1490 1500 1510 1520 1530 1540 1550 1560 1570 1580 1590 1600 1610 1620 1630 1640 1650 1660 1670 1680 1690 1700 1710 1720 1730

Fig. 6

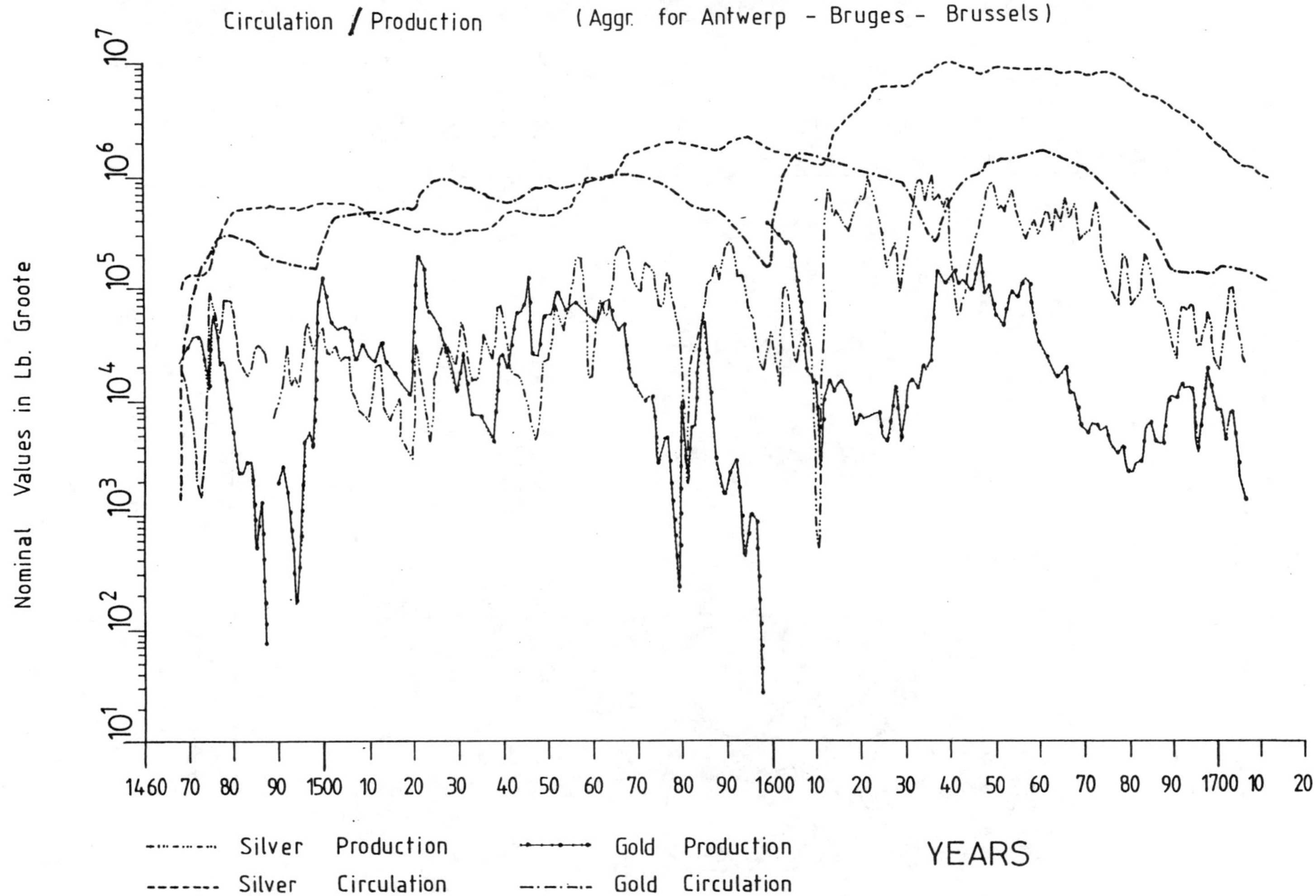
Circulation / Production
(Aggr. for Antwerp - Bruges - Brussels)
Nominal Values in Lb. Groote
10^7
10^6
10^5
10^4
10^3
10^2
10^1
1460 70 80 90 1500 10 20 30 40 50 60 70 80 90 1600 10 20 30 40 50 60 70 80 90 1700 10 20
YEARS
Silver Production
Silver Circulation
Gold Production
Gold Circulation

Fig. 7

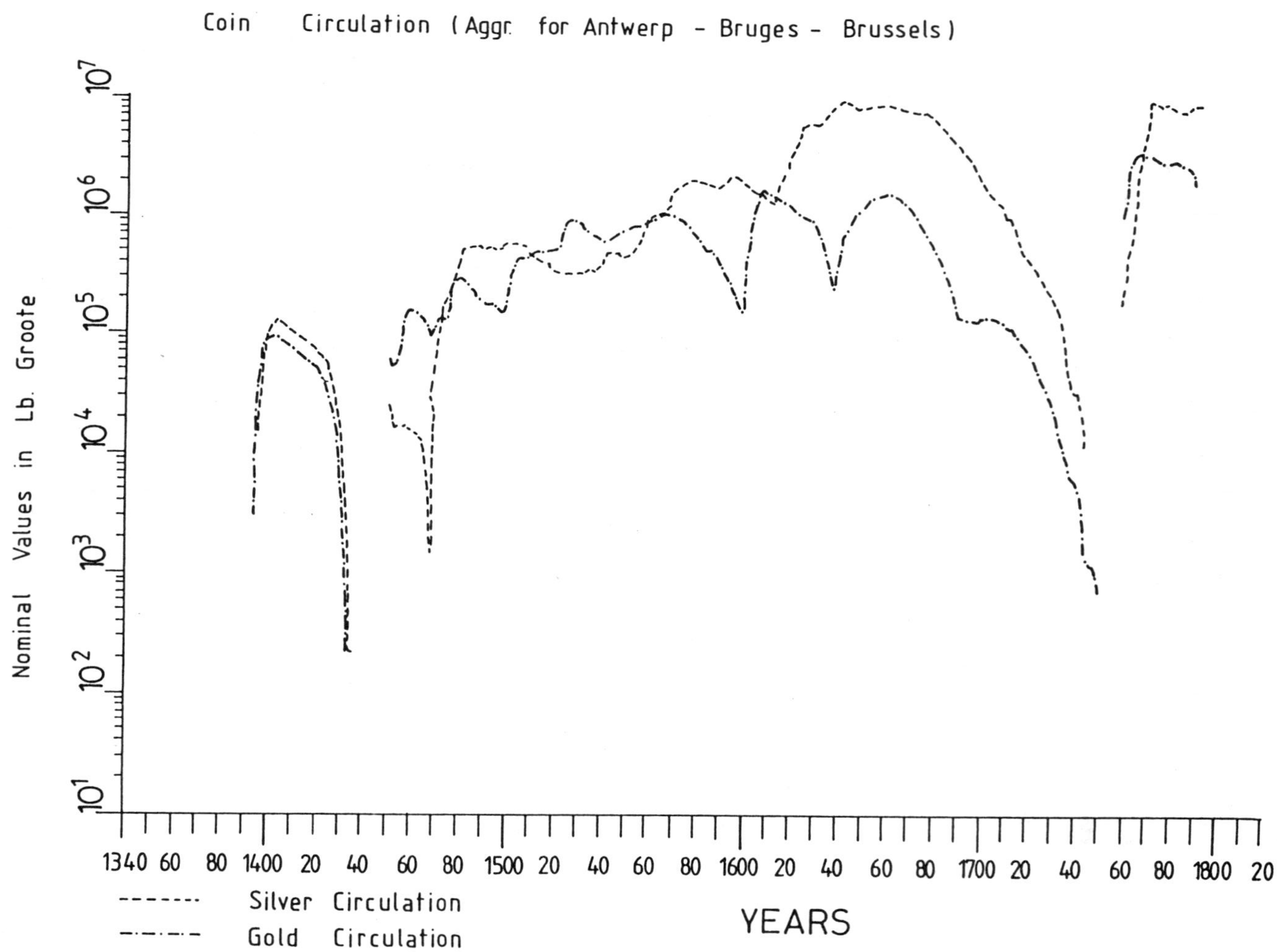
Coin Circulation (Aggr. for Antwerp - Bruges - Brussels)
Nominal Values in Lb. Groote
10^7
10^6
10^5
10^4
10^3
10^2
10^1
1340 60 80 1400 20 40 60 80 1500 20 40 60 80 1600 20 40 60 80 1700 20 40 60 80 1800 20
YEARS
Silver Circulation
Gold Circulation

Fig. 8

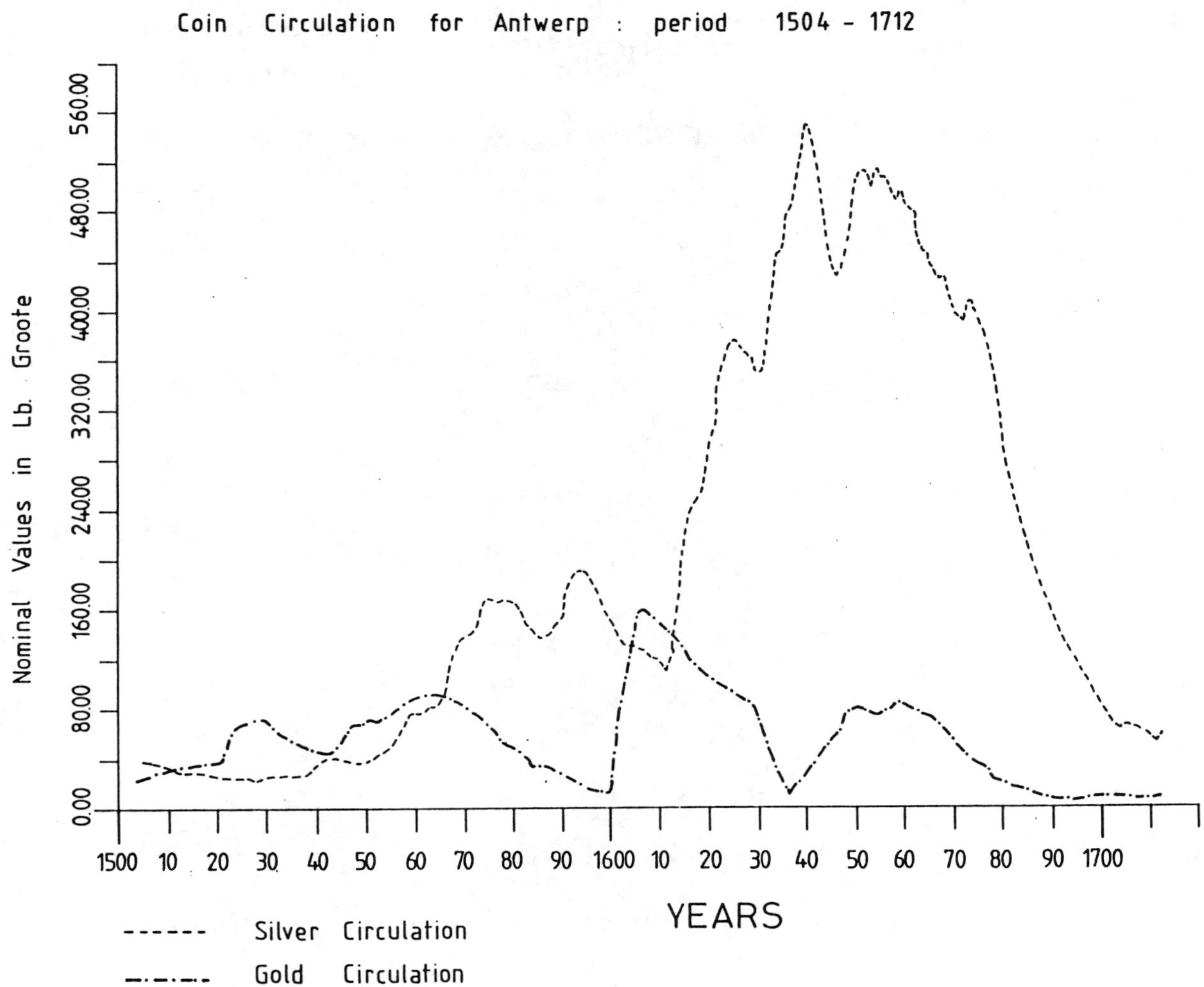

Coin Circulation for Antwerp : period 1504 - 1712
Nominal Values in Lb. Groote
0.00
80.00
160.00
240.00
320.00
400.00
480.00
560.00
1500
10
20
30
40
50
60
70
80
90
1600
10
20
30
40
50
60
70
80
90
1700
YEARS
Silver Circulation
Gold Circulation

Fig. 9

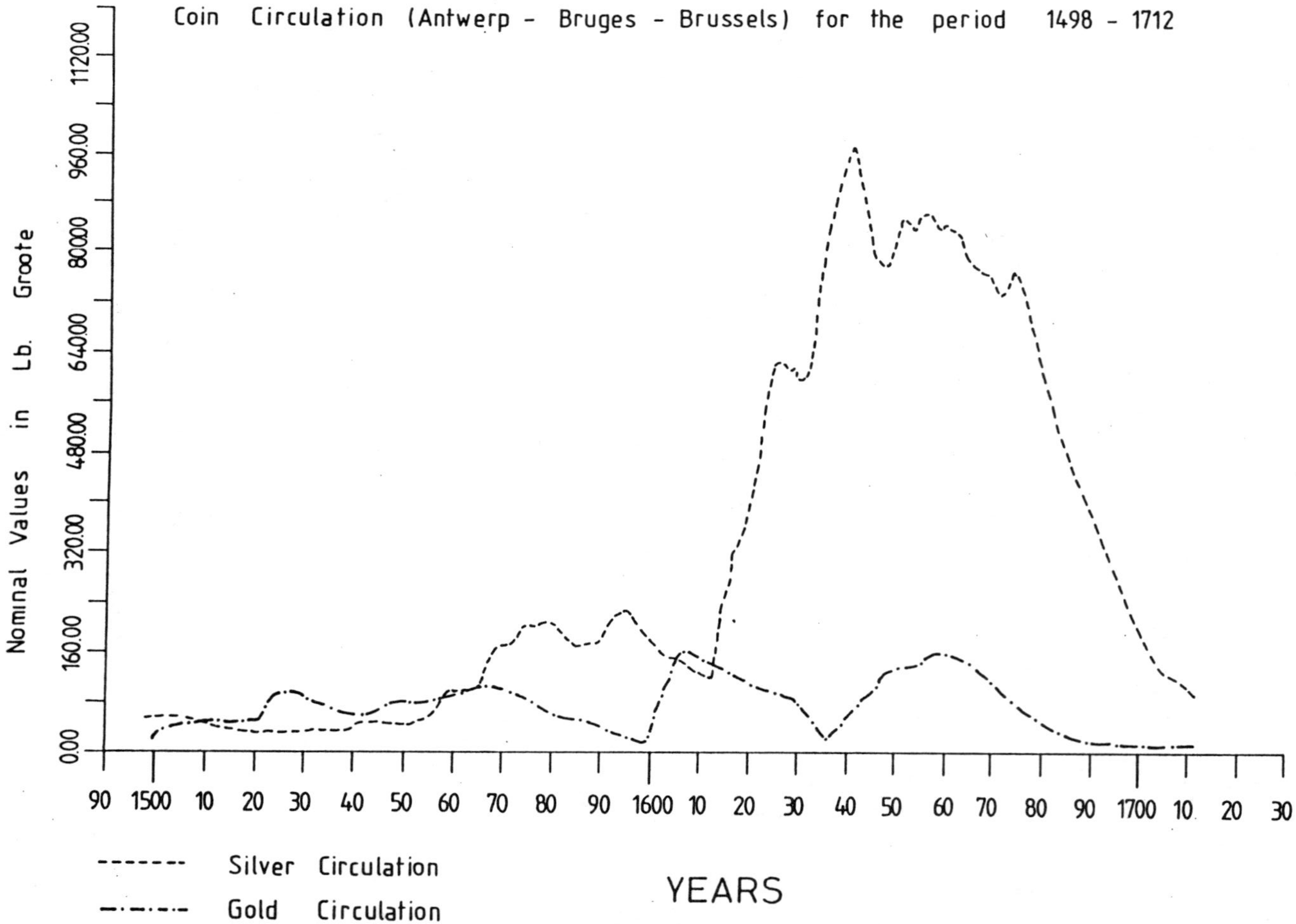
Coin Circulation (Antwerp - Bruges - Brussels) for the period 1498 - 1712
Nominal Values in Lb. Groote
0.00
160.00
320.00
480.00
640.00
800.00
960.00
1120.00
90 1500 10 20 30 40 50 60 70 80 90 1600 10 20 30 40 50 60 70 80 90 1700 10 20 30
YEARS
Silver Circulation
Gold Circulation

Fig. 10

PETTY COINAGE IN THE ECONOMY OF LATE-MEDIEVAL FLANDERS: SOME SOCIAL CONSIDERATIONS OF PUBLIC MINTING

by John H. MUNRO

The role of petty coinage in the monetary systems and economies of medieval Europe is one of the least well studied aspects of monetary history. The term «petty coinage» is itself vague enough to cause confusion. It does not mean, at least in this discussion, «token coinage»: i.e. base-metal jettons, of lead or copper, containing no silver, issued by private individuals or charitable organizations, unofficial coins lacking legal tender.[1] All of the coins discussed here were issued by the prince's mint according to officially prescribed monetary indentures; and all contained at least some silver, in amounts that diminished with debasements and rose with *renforcements*, respectively, more or less proportionately with changes in other related coins. All circulated as legal tender at official exchange rates, as some specified fraction of the silver penny or other «link» money anchoring the circulating coinage to the prevailing money-of-account. One modern synonym is thus «fractional coinage»; but such coinage was not necessarily «petty» everywhere in medieval Europe, in which relative coin values varied enormously. In medieval Flanders, the most common and apt contemporary term was «black money» - *zwart geld*, *monnaie noire* - because of the coins' very high base metal contents. That Flemish description thus accords with the modern numismatic term *billon*: an alloyed silver coin containing more than 50% copper. But that term, whose original medieval meaning was in fact «bullion», can be further misleading: in that many medieval coins with less than 50% silver were not considered to be petty; while others, despite high silver contents, such as English farthings, quarter-pence of sterling silver (92.5% fine), were.[2]

---

[1] See W.J. COURTENAY, *Token Coinage in the Administration of Poor Relief During the Later Middle Ages*, in: Journal of Interdisciplinary History 3 (1972–1973), pp. 175–195; Ph. GRIERSON, *Numismatics*, London, 1975, pp. 32–33 and 162–171.

[2] J.H. MUNRO, *Billon-Billoen-Billio. From Bullion to Base Coinage*, in: Revue belge de Philologie et d'Histoire 52 (1974), pp. 293–305. For the modern definitions of billon and black money, and their distinctions, see Ph. GRIERSON, *Numismatics*, p. 193.

In so far as modern-day scholars have discussed medieval petty coinage, we can find two quite contradictory views: either too much or too little was struck. Thus Van der Wee, in discussing the monetary history of late-medieval Flanders, has contended that frequently:

> «black money issues ... became excessive; under these circumstances an inflation of what we would call fiduciary currency occurred ... which resulted in a disappearance of the silver 'link money' from circulation, because of actual undervaluation».[3]

In terms of Gresham's Law, therefore, the overabundant and thus «cheap» or «bad» black money drove out the good «dear» silver money. For the directly contrary point of view, we can cite the opinion of Grierson that medieval:

> «moneyers preferred to strike high denominations to low ones ... and were apt to leave the public desperately short of small change».[4]

Much earlier, however, Cipolla had argued a third, more complex, but more positive view, combining elements of the other two arguments. On the one hand, he agreed that, by and large, medieval monetary authorities failed to ensure the proper functioning of a petty or fractional coinage system, for which two fundamental conditions were requisite: (1) that such petty coins be given a premium, i.e. a monetary or exchange value greater than their intrinsic commodity values; and (2) that the quantity of such coins be strictly limited in circulation in order to maintain that premium. Whenever such coins were given a higher intrinsic value approximating their monetary value, their issue «diminished drastically», because they were effectively undervalued (victims of Gresham's Law) and because their coinage was unprofitable for the moneyers. But debasements undertaken to reduce such intrinsic metal contents, to increase the premium (and the moneyer's profit) usually resulted in excess supplies in circulation, eventually eradicating the premium and halting production, until another debasement was undertaken. Nevertheless, if medieval mint policies seemed to promise no more than such absurd cycles, «a chain of alternate periods of shortages of petty coins and excessive coining», Cipolla came to the conclusion that, in medieval Italy at least, «the debasement of the petty coins and the failure to control their quantity did more good than harm»; i.e. by keeping the supply of petty coins

---

[3] H. VAN DER WEE, *International Business Finance and Monetary Policy in Western Europe (1384-1410)*, in: Business History Review 43 (1969), p. 375.

[4] Ph. GRIERSON, *Medieval Numismatics*, in: J. POWELL (ed.), Medieval Studies. An Introduction, Syracuse, 1976, p. 113; reprinted in Ph. GRIERSON, *Later Medieval Numismatics (11th-16th Centuries)*, London, 1979, n°. I; and Ph. GRIERSON, *Numismatics*, p. 97.

reasonably plentiful and elastic in an era when the «over-all supply of precious metals proved to be extremely inelastic».[5]

To understand fully the significance of this debate, and to see which view best reflects the historical reality of late-medieval Flanders, one must first realize that petty coins played a far greater role in medieval society than does «small change» in today's economy.[6] Certainly that was true of medieval Flanders: for, as Table 1 demonstrates, the purchasing power of such coins was surprisingly high, simply because prices (and wages) were then so low. The expression «a penny for your thoughts» takes on new meaning when we find, ca. 1350, that a silver penny *gros* would then buy 2.3 kg. of cheese and 3.6 kg. of wheat (4.8 1). Such a coin was hardly «small change» in this era.

A brief history of this silver *gros* and its related money of account will clarify the nature of petty coinage in late-medieval Flanders.[7] As a French royal fief, Flanders had used the prevailing monetary system of adjacent

---

[5] C. Cipolla, *Money, Prices and Civilization in the Mediterranean World, Fifth to Seventeenth Century*, Princeton, 1956 (reissued New York, 1967), pp. 27-33. Some of his arguments were based upon E. Cannan, *Money. Its Connection with Rising and Falling Prices*, London, 1918 (5th rev. edn. 1926), pp. 25-31, referred to also n. 34. For the concept of coinage 'premium' (implied in these arguments), see A. Feavearyear, *The Pound Sterling. A History of English Money*, 2nd rev. edn. by E.V. Morgan, Oxford, 1963, pp. 10-20; J.H. Munro, *Wool, Cloth, and Gold. The Struggle for Bullion in Anglo-Burgundian Trade (1340-1478)*, Brussels-Toronto, 1973, chapter 1, «The Economics of Bullionism,» pp. 11-42. See also C. Cipolla's related views in the important article *Currency Depreciation in Medieval Europe*, in: Economic History Review, 2nd ser. 15 (1963), pp. 413-422; reprinted, with a few changes, in S. Thrupp (ed.), Change in Medieval Society. Europe North of the Alps (1050-1500), New York, 1964, pp. 227-236; and also his recent *The Monetary Policy of Fourteenth-Century Florence*, Berkeley, 1982. For petty coinage in Venice, see F. Lane and R. Mueller, *Money and Banking in Medieval and Renaissance Venice*, I: *Coins and Moneys of Account*, Baltimore-London, 1985, pp. 10-11, 179-185, 201-203 and 258-259.

[6] See also C. Cipolla, *Money*, pp. 33-34; J.H. Munro, *The Purchasing Power of Coins and of Wages in the Low Countries and England (1500-1514)*, in: Sir R. Mynors, D. Thomson and W. Ferguson (eds.), The Correspondence of Erasmus, II: Letters 142 to 297(A.D. 1501-1514), Toronto, 1975, pp. 302-345.

[7] V. Gaillard (ed.), *Recherches sur les monnaies des comtes de Flandre*, I: *Depuis les temps les plus reculés, jusqu'au règne de Robert de Bethune*, Ghent, 1852, pp. 54-59, pp. 122-142 and doc. n°s. 1-15, pp. 3-29; H. Van Werveke, *Monnaie de compte et monnaie réelle*, in: Revue belge de Philologie et d'Histoire 13 (1934), pp. 126-133; H. Van Werveke, *Munt en politiek: de Frans-Vlaamse verhoudingen vóór en na 1300*, in: Bijdragen voor de geschiedenis der Nederlanden 8 (1953-1954), pp. 1-19; (both of which are reprinted in H. Van Werveke, *Miscellanea Mediaevalia*, Ghent, 1968, pp. 136-140 and 209-206.); A. Blanchet and A. Dieudonné, *Manuel de numismatique française*, II, Paris, 1916, pp. 115-240; E. Fournial, *Histoire monétaire de l'occident médiéval*, Paris, 1970, pp. 80-91; A. Grunzweig, *Les incidences internationales des mutations monétaires de Philippe le Bel*, in: Le Moyen âge, 59 (1953), pp. 117-172; R. Cazelles, *Quelques réflexions à propos des mutations de la monnaie royale française (1295-1360)*, in: Le Moyen âge 72 (1966), pp. 83-105.

northern France, the original royal system, based upon the denier and *livre parisis*, which were tied to the newer *tournois* system, inaugurated by Philip II Augustus (in 1205), by the fixed ratio of £ 1.00 parisis (20s.) = £ 1.25 tournois (25s.). In August 1266, Louis IX created a new heavy silver coin, the *gros*, of full *argent-le-roy* (95.83% fine), worth 12 current deniers tournois = 9.6d. parisis = 3d. English sterling (*esterlins*).[8] Then subsequently, in 1295, monetary stability in N.W. Europe abruptly ended with Philip IV the Fair's series of drastic debasements of both deniers tournois and parisis, causing the exchange value of the untouched gros to rise proportionately. Philip's debasements were in fact fiscal measures related to his wars with both Flanders and England. By 1319-1320, the Flemish had ousted the French armies and had won de facto political and monetary independence, though remaining de jure loyal subjects.[9]

At that time, with the return to stable money, the French gros circulated at 15d. tournois = 12d. parisis, as did the independently struck Flemish gros (issued from ca. 1300).[10] But fifteen years later, after a new round of French and Flemish debasements, their two coinages and moneys-of-account systems

---

[8] Commercially fine silver, known as argent-le-roy (AR), was 23/24ths (95.83%) pure: it consisted of 12 deniers, with 24 grains per denier. The gros tournois, struck by Louis IX in 1266 at a full 12 deniers argent-le-roy, 58 per marc de Troyes (244.753 g.), contained ideally 4.044 g. pure silver: 12.14 times as much as in the current denier tournois (0.333 g.: 220 struck to a marc of 3 deniers 18 gr.); and 9.75 times as much as in the denier parisis (0.415 g.: 212 to a marc of 4 deniers 12 gr.). The current English sterling penny, 243 struck to a Tower Pound (349.9144 g.) of 11 oz. 2 dwt. silver (92.5% pure), then contained theoretically 1.322 g. pure silver (see n. 17). The gros struck by Countess Margaret ca. 1275 evidently had only 2/3 the weight and value of the French gros.

[9] Philip IV did in fact debase the gros in August 1303: to 9 deniers AR, but restored it to a full 12 deniers AR in May 1305, and effected a complete reinforcement of the French coinage in September 1306. The next French debasement, a minor one, took place in January 1311. For the political and economic events, see D.P. Blok, W. Prevenier et al. (eds.), *Algemene geschiedenis der Nederlanden*, II: *Middeleeuwen*, Haarlem, 1982; H. Pirenne, *Histoire de Belgique*, I, 5th edn. 1928 (new illustrated edn. Brussels, 1956), pp. 225-256; J. Lestocquoy, *Les villes de Flandres et d'Italie sous le gouvernement des patriciens, XIe - XVe siècle*, Paris, 1952, pp. 133-139.

[10] In March 1318, the French gros tournois underwent a slight reduction in its weight and silver content (taille increased from 58 to 59 1/6 per marc), from 4.044 g. to 3.964 g. silver. Meanwhile, or in the 1320s, the Flemish gros (groot) was evidently being struck at 57 per marc of 11 deniers AR, for a pure silver content of 3.772 g. = 95.2% of that in the current French gros. (By ordinance of April 1300, the Flemish gros had then contained 3.794 g. pure silver: 56 1/2 per Cologne marc (233.856 g.) of 11 den. 11 1/2 gr. AR). See the sources by A. Blanchet, A. Dieudonné, E. Fournial and V. Gaillard cited in n. 7 above; and F. and W. Blockmans, *Devaluation, Coinage, and Seignorage under Louis de Nevers and Louis de Male, Counts of Flanders (1330-1384)*, in: Coinage in the Low Countries (880-1500), (British Archeological Reports, International Series 54), Oxford, 1979, pp. 70, 83 (Table 2).

permanently parted company. While the (stable) French gros once more rose in value, in terms of debased deniers tournois and parisis, the Flemish gros coins, no longer French, continued to maintain a relationship, now permanently frozen, of 1d. Flemish gros = 12d. Flemish parisis (or 1 shilling). In fact, the Flemish parisis system had really lost its own coins to become instead simply a money-of-account subsidiary to the Flemish gros system, whose silver penny coin had become the 'link money' or anchor of the following monetary relationships prevailing until the French Revolution:[11]

1 livre gros [*pond groot*] = 12 livres parisis
= 240s. par.
1 livre gros = 20s. = 240d. gros
1s. gros [sol, sou, shilling]
= 12d. = 36 esterlins = 288 mites
1d. gros = 3 esterlins [*ingelschen*, sterlings]
= 24 mites
1d. gros = 1 sol parisis = 12d. parisis
1d. parisis = 2 mites gros

For many years, the largest silver coin struck was the penny gros itself: until the inflationary 1360s, when Count Louis de Male, after having debased the penny gros to just under 40% of its original (1300) silver content and value, introduced the first double gros (2d).[12] The other Flemish silver coins struck in the fourteenth century were thus subdivisions of the penny: the demi-gros (1/2 d), the esterlins (1/3d - also known as *ingelschen* and *tierchelets*: 1337-1338 and 1351-1352, only); quarter-gros (1334-1336 only); and mites (*mijten*), first single and then double mites.

[11] See sources cited in n. 7, and also J. GHYSSENS, *Le monnayage d'argent en Flandre, Hainaut, et Brabant au début de la Guerre de Cent Ans*, in: Revue belge de Numismatique 120 (1974), pp. 109-191; H. VAN DER WEE, *The Growth of the Antwerp Market and the European Economy (Fourteenth-Sixteenth Centuries)*, I: *Statistics*, The Hague, 1963, pp. 107-135.

[12] The silver lion or Botdragher: 57 per marc of 8 deniers AR, containing 2.743 g. pure silver, first struck by Louis de Male at Ghent, from 12 April 1365. The previous (single) gros, struck at Mechelen from 1 Dec. 1363 to 2 Mar. 1364, at 78 per marc of 6 deniers AR, contained 1.503g. pure silver (vs. 3.794 g. in the 1300 gros). See H. VAN WERVEKE, *De muntslag in Vlaanderen onder Lodewijk van Male*, (Mededelingen van de Koninklijke Vlaamse Academie voor Wetenschappen, Letteren, en Schone Kunsten, Klasse der Letteren 11:5), Brussels, 1949, account n°. ix, pp. 20-21; V. GAILLARD (ed.), *Recherches sur les monnaies des comtes de Flandre*, II: *Sous les règnes de Louis de Crécy et de Louis de Male*, Ghent, 1856, doc. nos. 84-85, pp. 173-176; no. 87, pp. 179-180. See also Ibid., I, no. 14, pp. 27-28 (n. 7 above). On the debasements and inflation of the 1360s, see J.H. MUNRO, *Mint Outputs, Money, and Prices in Late-Medieval England and the Low Countries*, in: E. VAN CAUWENBERGHE and F. IRSIGLER (eds.), Münzprägung, Geldumlauf und Wechselkurse. Minting, Monetary Circulation and Exchange Rates, (Trierer Historische Forschungen 7), Trier, 1984, pp. 31-32, 49-52, and Tables A.8, B.1-7, and Graphs 1-10, pp. 94-118.

These last coins, the Flemish mites, are indeed the true *monnaie noire*, *zwart geld*, and the chief petty coins of this discussion.[13] When their minting is first recorded, in October 1334, their fineness was 1 denier 18 grains out of 12 deniers argent-le-roy (14.58%); and each contained 0.130 gram pure silver, just 3.73% of the current gros' silver content.[14] By the 1380s, the silver content of the mite (double) had been reduced by over 70% to just 12 grains argent-le-roy, or 4.17% commercially fine (= 0.041 g. in a coin weighing 1.02 g.).[15] These mites may be considered as the vestigial physical remnants of the old parisis system, with the double mite, known as the *courte*, valued at 1/12th gros and thus at 1d. parisis. They are the only form of Flemish petty coinage up until 1410, when the quarter-gros reappears in much debased form as the *gigot* or *zeskin* (= 6 mites), at only 4 deniers argent-le-roy, or 33.3% fine.[16] It thus qualified as billon by the modern numismatic definition, but was always regarded as «silver» rather than as «black» money in the Flemish mint accounts.

These Flemish mint accounts, the most complete such accounts extant for late-medieval Europe, virtually continuous from 1334, permit us to answer the question implicitly posed earlier: was too much or too little petty coinage struck in this era? To be sure, the English mint accounts go back a century further, to 1234, and are also quite continuous; but they do not provide the denominations of the silver coins struck after 1351.[17] Furthermore, the peculiar nature of the English coinage makes it unsuitable for this particular study: for, with the exception of the years 1335-1344, the English coinage, even in halfpence and farthings (1/4d) - was always struck at sterling fineness, 11 oz. 2 dwt or 92.5% pure silver.[18] Changes in the base metal content of

[13] See J. Ghyssens, *Les débuts de la mite et de la monnaie de compte en Flandre*, in : Revue belge de Numismatique 116 (1970), pp. 181-189; A. Dieudonné, *Les doubles cornus de Philippe le Bel et les origines de la mite flamande*, in: Revue belge de Numismatique 85 (1933), pp. 23-41.

[14] These are probably single mites (264 per marc). V. Gaillard, *Recherches*, II, n°. 17, p. 34. See Table 2. The current gros, or rather two demi-gros, then contained 3.484 g. pure silver (1332-1337).

[15] See Table 2, and Algemeen Rijksarchief (ARA), Rekenkamer, Rolrekening n°. 824 (Feb.-Aug. 1389).

[16] See Table 4; and note that the original quarter-gros of 1334-1336 had enjoyed a fineness of 85.4% AR.

[17] See C.E. Blunt and J.D. Brand, *Mint Output of Henry III*, in: British Numismatic Journal 39 (1970), pp. 61-66; C.G. Crump and C. Johnson, *Tables of Bullion Coined Under Edward I, II, and III*, in: Numismatic Chronicle 4th ser. 13 (1913), pp. 200-245; G.C. Brooke and E. Stokes, *Tables of Bullion Coined from 1377 to 1550*, in: Numismatic Chronicle 5th ser. 9 (1929), pp. 27-69; Sir J. Craig, *The Mint. A History of the London Mint from A.D. 287 to 1948*, Cambridge, 1953; J.H. Munro, *Bullion Flows and Monetary Contraction in Late-Medieval England and the Low Countries*, in: J.F. Richards (ed.), Precious Metals in the Later Medieval and Early Modern Worlds, Durham, 1983, pp. 97-158.

[18] The fineness of halfpence and farthings only (no pence were struck) was reduced from 11 oz. 2 dwt. to 10 oz silver (83.3% pure). See sources cited in n. 17.

the Flemish silver coinage, from mites to gros, are essential to the role of petty coinage in that county.

There are other important reasons for focusing upon late-medieval Flanders, and indeed upon that era (1334–1484 in this study). In the first place, Flanders was the most urbanized, commercialized, industrialized, monetized, and economically advanced region in late-medieval Europe north of the Alps. Second, this 150 year period encompasses most of the late-medieval secular economic decline or contraction (sometimes called the medieval «Great Depression»). Certainly Flanders itself periodically suffered from agrarian, commercial, and industrial malaise in this era; and the mainstays of her economy, the textile industries and trades, were being undermined by serious competition, in contracting markets, from draperies in neighbouring Brabant, Holland, and especially England. Furthermore, an important aspect of this late-medieval secular economic decline, according to some historians at least, was a severe monetary contraction with strongly deflationary tendencies, both of which can be well documented for Flanders and the adjacent Low Countries. At the same time, this was also an era of intensive, widespread warfare, most notably the Hundred Years' War, intimately involving Flanders on several occasions; and that warfare was in part financed by coinage debasements that in turn spawned veritable *guerres monétaires* and other measures in an international competition for scarce bullion. Both the warfare itself and the accompanying coinage debasements were highly inflationary, overpowering and temporarily reversing the deflationary forces of the era. But the latter forces reasserted themselves during periods of peace, which usually also brought about coinage renforcements, themselves deflationary (by converting current debased coins into fewer «good» coins).[19] While England had relatively few such debasements (1335, 1344–1351, 1411, 1464), Flanders (and dependencies) had a great many: 32 of the silver coinage from 1334 to 1484, in oscillating cycles: (a) 1334–1389: warfare, debasement, and inflation; (b) 1390–1416: relative peace, monetary reform, strong coinage, and deflation; (c) 1416–1433: warfare, debasement, severe inflation; (d) 1433–1474: relative peace, monetary reform, strong coinage (altered 1466), and severe deflation; (e) 1474–1480s (to 1496): warfare of Charles the Bold and then of the Habsburg Maximilian, more debasements, and severe inflation, coinciding with the influx of new silver from Central Europe.[20] During

[19] See evidence cited in J.H. MUNRO, *Bullion Flows*, pp. 97-158; ID., *Mint Outputs, Money and Prices*, pp. 31-122 (see n. 12); ID., *Wool, Cloth, and Gold* (see n. 5); ID., *Monetary Contraction and Industrial Change in the Late Medieval Low Countries (1335-1500)*, in: N. MAYHEW (ed.) Coinage in the Low Countries (880-1500), Oxford, 1979, pp. 95–161; J.H. MUNRO, *Economic Depression and the Arts in the Fifteenth-Century Low Countries*, in: Renaissance and Reformation 19 (1983), pp. 235-250; and H. VAN DER WEE, *Antwerp Market*, II, pp. 7-118 (see n. 11).

[20] See the sources cited in nn. 7, 11, 12, 17, 19 and 41–46.

such periods of inflation, one might suppose that the relative importance of petty coinage would diminish: in that the need for higher denomination coins would grow with rising prices, while that for low denomination coins would fall. Furthermore, with debasements, the costs of issuing petty coinage would rise: partly from the additions of more base metal (increasing the difficulty of achieving the prescribed fineness), but more in the extra labour required in striking a greater number of coins to the *marc* (mint-weight unit). The converse would be true with reinforcements; and from Table 2, one can see (final column) that the total mintage fees as a proportion of the total value of coins struck per marc of silver rose with debasement and fell with reinforcement. With the ensuing deflations, exacerbated by other (above-mentioned) factors, the demand for lower value, petty coins would rise, along with their increased purchasing power (as noted in Table 1). To what extent these suppositions are borne out will be determined in the ensuing analysis.

From the remaining Tables 3-5, based on the Flemish mint accounts we can determine the actual quantities of petty coinage struck, and the proportions of total coinage (and total pure silver) struck in that form from 1334 to 1484. Table 5 provides a comparison between the silver contents of the petty and gros coinages over this entire period. In summary, we note that the Flemish mint accounts record the striking of petty coins (mites and from 1410, quarter-gros) for only 59 or 39% of these years. Although the coinage of mites is recorded in the very first extant Flemish mint accounts, October 1334 - November 1338, no more mites are struck (or recorded) for another 36 years, not until January 1375, when the first double mites were issued. These in turn were struck only up to June 1377; their coinage is not resumed for another 12 years, in February 1389, ten months before the monetary reform of December 1389. With that reinforcement, new double mites (with 29% more silver) were issued, and ceased only with the (temporary) closure of the Flemish mints in June 1402.[21] Although the mints were reactivated in January 1410, and coinage of quarter-gros (*gigots*) was resumed, no more mites were struck until late 1419, 15 months after the commencement (June 1418) of another 15-year cycle of frenetic coinage debasements, under Philip the Good.

With the Burgundian monetary unification, based on the reformed, strengthened Flemish coinage, of 1433-1434, a new issue of double mites was

[21] See V. Gaillard, *Recherches*, II, n° 102, p. 204; ARA, Rek. Rol. n° 824-826 (Ghent mint); n° 776-787 (Bruges mint); *Rekenkamer*, registers n° 48, 976-977 (Mechelen mint). None was struck at the Falkenberg (Fauquemont) mint. See also J.H. Munro, *Mint Policies, Ratios, and Outputs in England and the Low Countries (1335-1420)*, in: Numismatic Chronicle 8th ser. 1 (1981), pp. 71-116; H. Van Werveke, *De Vlaamse Munthervorming van 1389-1390*, in: Nederlandsche Historiebladen 1 (1938), pp. 336-347 (reprinted in his Miscellanea Mediaevalia, Ghent, 1968, pp. 268-280).

struck, with 25% more silver than the 1430–1432 issues had contained.[22] After a quarter-century of virtually continuous if much diminishing output, that issue was briefly replaced by another in 1457, reduced by 10% (in weight alone); and that issue ceased the following year when the Flemish mints, starved for bullion, were coining only a trickle of metal.[23] Philip the Good, nearing the end of his reign, reactivated the mints with a mild debasement in June 1466, assisted by the first arrivals of the new South German silver. From that time, issues of mites and quarter-gros are more or less continuous into the sixteenth century. The only significant exception, when none is minted, is the period April 1474 to December 1477, after the 1474 monetary ordinance had in fact pointedly refrained from prescribing the issue of quarter gros and mites.[24]

We may now analyze the entire 150 year period, with the aid of tables 2–5. First, in terms of the proportion of total silver bullion minted (marcs argent-le-roy), and taking 10–year periods from the mid–1330s, the amounts struck as petty coins were rarely more than 1%, and were generally under 1%, until the mid-fifteenth century. The same is also true when viewed in terms of the total money-of-account values of the silver coinages struck, if very slightly greater, to take account of the higher mintage fees and thus the higher *traite* per marc for the petty coinage. Furthermore, if we exclude those quarter-gros struck from 1410, and focus only on the mites, the true monnaies noires, then those proportions of total coinage struck in this form drop dramatically - to about 1/10th of 1%, with the significant exception of the mid fifteenth century, to be considered later as a separate issue.

If, however, we consider the outputs of petty coinage in terms of the number of coins struck (third set of columns in Table 3), we see that in some decades between 3 and 4.5 millions of such petty coins were struck, accounting for as much as 18% of the total number of coins issued, again until that anomalous mid-fifteenth century period. If the prince and his mint-masters had been questioned about meeting their social responsibilities in providing adequate amounts of small coinage, undoubtedly they would have pointed to such figures - perhaps with some justification. But many other observers, if enabled to view this 150–year span of statistics, would undoubtedly echo the

[22] See Tables 2-3; ARA, Acquits de Lille, liasse n° 937; J.H. Munro, *Wool, Cloth, and Gold*, pp. 98–103; P. Spufford, *Monetary Problems and Policies in the Burgundian Netherlands (1433-1496),* Leiden, 1970, pp. 4-6, 30-31, 44-46; Minting of the new, unified Burgundian coinage commenced in Flanders (Ghent) on 15 Oct. 1433 in Hainaut, on 26 Nov. 1433; in Holland, on 23 Jan. 1434; in Brabant, not until 4 Sept. 1434.

[23] See nn. 42–45.

[24] See Tables 2-3; ARA, Rek. reg. n° 580, f° 100^{v} (1474); reg. n° 18, 110 (Bruges mint account, 20 Dec. 1474 - 14 Mar. 1476); L. Deschamps De Pas, *Essai sur l'histoire monétaire des comtes de Flandre de la maison de Bourgogne*, in: Revue Numismatique 2nd ser. 7 (1862), pp. 357-360 (text of ordinance of 27 Oct. 1474).

criticism of Grierson: that remarkably little, indeed scandalously little, silver bullion went into the minting of these petty coins, the coins that the bulk of the public, whether peasants or urban artisans, needed for their daily bread and drink.

In particular, one must first ask why no petty coins at all were recorded in the mint accounts from 1338 to 1375. Two possible reasons can be suggested. First, the earliest extant mint accounts (1334–1338) indicate that the count (Louis de Nevers) had exacted no seignorage, no profit at all, on the coining of mites; only brassage was exacted, to cover the cost of the copper and the mint master's other expenses (and his profit, presumably).[25] Since these mint accounts are essentially financial reports to the count setting down and explaining the seignorage revenues, subsequent mint-masters may have seen no need to record the striking of any mites, if that policy of coining mites free of seignorage had continued until the 1370s, when the next issue of mites is recorded (January 1375); and, as Table 2 shows, the coinage of those double mites was indeed subjected to seignorage.[26] An alternative but related explanation (hypothesis) is that, if no seignorage had been charged on the coining of mites up to January 1375, then the count would have had little incentive to require that any such coins be struck. The successor to Count Louis de Nevers, we must remember, was Louis II de Male, ill-famed for his extravagant debasements and other mint-manipulations (gold: silver ratios) to generate very imposing mint profits (1346–1384), as Van Werveke has shown. The Flemish coinage was reformed and strengthened only under the regime of his son-in-law successor, Duke Philip the Bold of Burgundy.[27] But possibly a better answer may be found in the more general hypotheses to explain why so little petty coinage was struck in this 150 year period (excepting the 1440s and 1450s).

The chief explanation offered by Grierson is that the costs of minting petty coinage or monnaies noires were much too high to be profitable or even acceptable to the mint-masters, who certainly exerted a strong influence on the prince's monetary ordinances (generally in medieval Europe). As Grierson notes, «twelve times as much labour was involved in making 12 pennies as in making one shilling,» a valid point, if we are willing to consider the French and Flemish gros (and some Italian *grossi*) as shillings.[28] That last

[25] V. GAILLARD, *Recherches*, II, n° 17–21, 31–58: «De le noire monnoye, c'on appielle mites; somme de noire monnoie 2,637 marcs, dont je ne fai nulle compte, pour che que Monseigneur ne prent riens de seignorage» (p. 45).

[26] Ibid., II, n° 102, p. 204.

[27] H. VAN WERVEKE, *Currency Manipulation in the Middle Ages. The Case of Louis de Male, Count of Flanders*, in: Transactions of the Royal Historical Society 4th ser. 31 (1949), pp. 115–127; (reprinted in his Miscellanea Mediaevalia, pp. 258–67); H. VAN WERVEKE, *Muntslag* (see n. 11); F. and W. BLOCKMANS, *Devaluation, Coinage, and Seignorage*, pp. 69–94 (see n. 10); J.H. MUNRO, *Mint Policies*, pp. 71–116 (see n. 21).

[28] Ph. GRIERSON, *Medieval Numismatics*, p. 113 (n. 4); ID., *Numismatics*, p. 97 (n. 1); P. SPUFFORD, *Monetary Problems*, pp. 44–46 gives a similar argument for the

argument, however, really pertains only to the peculiar structure of the English medieval coinage, when halfpence and farthings had precisely the same sterling fineness as pence and groats, had exactly proportional weights, 50% and 25% of the penny's weight respectively, and bore proportionally the same brassage and seignorage charges as the pence. That meant that, per Tower Pound of sterling silver, farthings, halfpence, and pence all had the same traite or money-of-account values.[29] But in the Flemish monetary system, the combination of reductions in both weight and fineness, permitting the possibility of proportionally more (or fewer) coins struck and/or more copper alloy per marc, meant that the petty coins contained proportionally much less silver (a double mite less than 1/12th) than contained in the gros, as Table 5 shows. Consequently the petty coins enjoyed a far higher traite per marc de Troyes (i.e. value of the coinage so struck), thus permitting the exaction of a far larger brassage fee (and seignorage) to cover both the raw material and labour costs fully - and allow some profit to the mint-master. Merchants and others receiving such petty coins from the mint and subsequently in trade were presumably willing to accept them with proportionately smaller silver contents (than in the gros) because of their higher utility in effecting day to day payments, buying food and necessities. So long as the total of mintage fees on monnaies noires did not exceed the real utility to those spending them, there would be no objection to the coins' composition.

Consequently, we cannot assume that the coinage of monnaies noires was necessarily a losing proposition for the mint-master (or the prince).[30] Instead, the real question to be asked is this: were the brassage fees that the prince authorized in each mint indenture always sufficiently high to ensure that the mint-master would indeed cover all his manufacturing costs and receive a fair profit? Thus one can discern in the 1466-1467 monetary ordinances (Table 2) a sharp reduction in these brassage fees, despite a general debasement of silver coinages effected by these ordinances. That reduction, even though designed to increase the mint price for merchants' bullion, may explain why relatively little petty coinage was minted thereafter (and, as noted before, none at all in 1474-1477: see Tables 3 and 5).

An alternative, though quite complementary explanation is that sometimes the prince himself exacted too high a seignorage fee at the expense of the mint's price offered for bullion or at the expense of mint-master's brassage (or both). In some instances, that complaint may well have been true. Thus

---

Burgundian Low Countries; C. Cipolla, *Money*, p. 28 (n. 5) implicitly supports this view (under the conditions specified above). F. Lane and R. Mueller, however, found that «Venice began to make very large profits on black money before the end of the fourteenth century»: *Money and Banking*, I, p. 201-202 (see n. 5).

[29] See n. 18 Ph.; Grierson, *Numismatics*, p. 33.

[30] See F. Lane and R. Mueller in n. 28 above.

when the (recorded) coinage of mites was resumed in January 1375, the rapacious Count Louis de Male exacted a seignorage of 12d per marc argent-le-roy on double mites, six times that charged (2d) on the coinage of double gros. In November 1376, he was forced to reduce the seignorage on mites to 8d, ceasing their coinage in June 1377. Subsequently and similarly, Duke Philip the Bold's coinage of mites in 1389 bore a seignorage charge of 12d. per marc, double that charged on the coinage of gros and double gros (6d).[31] Whether such hefty seignorage fees were at the expense of the mint price for bullion or of the brassage cannot be ascertained for certain. The later Burgundian mint indentures provide more explicit detail, and the one dated October 1467, presented in Table 6, does show that the Flemish mints then offered merchants a higher price for bullion to be coined into double patards (4d gros) than for bullion struck into double mites: 27s. 4d. vs. 26s. 11d. gros. Two important qualifications, however, must be noted: the lower mint price for bullion pertained to all coins from the patard (double gros) on down to the mites - the double patard was the sole exception enjoying the higher mint price; and the seignorage per marc was the same on all coins. Furthermore, we must recall that the issue of mites in 1334-1338 was free of seignorage; and also note that from 1390 to 1402 the seignorage on mites was only half of that charged on the coinage of gros (2d vs 4d: see Table 5). Thereafter, the Burgundian dukes exacted the same seignorage on all silver coinages, with the sole and brief exception of the years 1430-1432, when the seignorage charged on coining mites was 1d. higher than on other coins. Our conclusion must be that excessive seignorages or insufficient brassages can satisfactorily explain the low volumes of petty coinage outputs only in very few instances.

For another related explanation, we should reconsider the observations of Cipolla, cited earlier, on the requirements for an effective petty coinage.[32] Perhaps, therefore, the Flemish counts and their Burgundian ducal successors (from 1384) deliberately restricted the issues of petty coinage, more so than other medieval princes, precisely in order to avoid those perils to which Cipolla and Van der Wee alluded. Certainly there is abundant evidence that fourteenth-century Flemish monetary authorities were well aware of the anachronistically named «Gresham's Law», the version in Nicolas Oresme's treatise *De Moneta*, that cheap money drives out dear.[33] But there is no

[31] See Tables 2 and 5, and nn. 25-26 above.

[32] See n. 5.

[33] See J.H. Munro, *Wool, Cloth, and Gold*, chapters 1-4 (see n. 5); Id., *Mint Policies*, pp. 71-116 (see n. 21); H. Laurent, *La loi de Gresham au moyen âge. Essai sur la circulation monétaire entre la Flandre et le Brabant à la fin du XIVe siècle*, Brussels, 1933, with corrections by J.H. Munro, *Monnayage, monnaies de compte et mutations monétaires au Brabant à la fin du moyen age*, in: J. Day (ed.), Etudes d'histoire monétaire XII-XIXe siècles, Lille, 1984, pp. 263-294; Ch. Johnson (ed.), *The De Moneta of Nicholas Oresme and English Mint Documents*, Londen, 1956, especially

explicit evidence whatsoever that such considerations ever entered into decisions on how much petty coinage was to be struck. Furthermore, as Cannan (on whose treatise Cipolla had drawn) had correctly stated, the mintage fees imposed (brassage and seignorage), as a cost to those furnishing bullion, will restrict production, necessarily limit the supply of petty coins and thus «will tend to raise the ordinary [exchange] value of the coin above that of the uncoined metal by the amount of the charge».[34] But that observation is in fact true of all coinage, not just petty coins. For the conversion of bullion into coin can never be costless, and can be undertaken only so long as the current coin commands a premium over bullion sufficient to cover those minting costs. Normally coins of all denominations will command such a premium, because transacting trade and effecting payments is so much more convenient in legal tender coin - coins bearing the prince's stamp attesting to their fineness, weight, precious metal contents, and exchange values - than in raw bullion, which necessarily has to be assayed and weighed to ascertain its market value. Laws forbidding payment or trade in bullion and foreign coin, requiring such metal to be delivered to the mint, so far as they were enforced, served to increase the premium on coin over bullion. If that premium ever fell below the sum of brassage and seignorage, bullion would cease to flow to the mint: requiring the prince either to reduce the mintage fees or to increase the traite value of bullion, the mint price, and thus the coinage premium by a debasement.[35] All of these monetary relationships, including the mintage charges as percentages of the coinage traites, for both the petty and gros coinages, during cycles of both debasements and renforcements, from 1384 to 1484, can be seen in Tables 2 and 5. Note in particular how much higher, absolutely and relatively, were the mintage fees on the petty coinage; but recall the earlier observation that such charges were justified not only by real production costs but more particularly by the coins' specific utility and hence the demand for them in local circulation.

Indeed, we should consider the demand for coinage on the part of those merchants and money-changers who furnished bullion and dealt with the mints: for their demands would determine the composition of the coinage struck. Of course, for reasons already elaborated, there was also a public demand for petty coinage. But to what extent was that public demand reflected by those merchants dealing directly with the mint? To the extent that such

---

chapter xx, pp. 32-33. Laurent and others, however, have suggested that the critical passages were additions by 14th century Flemish mint officials: H. LAURENT, *Le problème des traductions françaises du Traité des monnaies d'Oresme dans les Pays-Bas bourguignons*, in: Revue d'Histoire Economique et Sociale 21 (1933); Ch. JOHNSON, Ibid, p. 7; E. BRIDREY, *La théorie de la monnaie au XIVe siècle*, Paris, 1906.

[34] E. CANNAN, *Money*, pp. 25-26 (see n. 5).

[35] See n. 5 (A. FEAVEARYEAR and J.H. MUNRO).

merchants were engaged in larger-scale forms of trade and finance, involving the Low Countries' regional and international trades, whose importance for Flanders has already been stressed, we suppose that the monetary demands of such merchants and money-changers would be highly biased in favour of larger denomination coins. Consider, furthermore, the different spheres of circulation for petty and large-denomination coins, silver and gold. The circulation of the petty Flemish coinage, certainly the monnaies noires, was strictly local, just within Flanders itself; but the higher denomination silver gros and patards had a much wider circulation - regionally widespread throughout the Low Countries and the Rhineland, and further abroad, internationally. Proof of that can be seen in the statistics of the Calais mint for the Calais Staple wool trade. When the relative Anglo-Burgundian mint ratios favoured silver over gold at Calais, with Flemish (Brabantine and Dutch) purchases of English wool at the Staple there, almost all the payments (from 1427 to ca. 1440) were made in Flemish silver.[36]

The first of the two major hypotheses to be put forward, therefore is that the production of petty coinage was largely in response to the mercantile demands of those dealing with the mints, and was a reflection in particular of the composition of that demand. Given Flanders' very prominent role in both regional and international trade, one major reason why so much, indeed almost all, silver bullion brought to Flemish mints was struck in high denomination coins was mercantile demand for such silver coins in servicing both domestic and foreign trade.

Questions of demand cannot, of course, be divorced from those of supply: in particular, one involving those differences in coinage circulation just discussed that has an equally strong bearing on the composition of the Flemish silver coinage outputs. Mint production is to some extent a question of *replenishing* existing monetary stocks currently required by the economy; and those stocks are not imperishable. They diminish from wear and tear in coinage circulation; from precious-metal exports in financing international trade, warfare, and diplomacy; from piracy, shipwreck, brigandage, and simple loss; and from hoarding (of metal never retrieved).[37] To be sure, precious-metal outflows - as distinct from permanent losses - are often offset

[36] J.H. MUNRO, *Wool, Cloth and Gold*, chapter 4-5, especially Table 1, p. 95; Graph IV, p. 97 (see n. 5). P.R.O., L.T.R., E. 364/59, 61-3, 65-6, 69; 72; K.R., E.101/192-3; Calendar of Patent Rolls, 1422-1429, pp. 337-338; CPR 1429-1436, pp. 256-259.

[37] See some different views on this in C.C. PATTERSON, *Silver Stocks and Losses in Ancient and Medieval Times*, in: Economic History Review 2nd ser. 25 (1972), pp. 205-235; N. MAYHEW, *Numismatic Evidence and Falling Prices in the Fourteenth Century*, in: Economic History Review 2nd ser. 27 (1974), pp. 1-15; J.H. MUNRO, *Bullion Flows*, pp. 98-102 (see n. 17); Ph. GRIERSON, *Numismatics*, pp. 146-149 (see n. 1); ID., *Medieval Numismatics*, p. 125 (see n. 4); ID., *Coin Wear and the Frequency Table*, in: Numismatic Chronicle 7th ser. 3 (1963), pp. 1-13, (reprinted in his *Later Medieval Numismatics*, London, 1979, n° 19).

by metal inflows; but such inflows are normally reflected in subsequent mint outputs, outputs of predominantly high denomination coins. By and large coinage outflows were also largely in high denomination coins, with high precious metal contents, for obvious reasons. Petty coins, especially monnaies noires, were not so subject to these outflows. For their very low silver and high base metal contents, and thus their high mintage fees almost fully protected them, and Flanders, against export as bullion, or as coin for circulation abroad.[38] Similarly, the monnaies noires were much less subject than good silver coin to other forms of loss, domestic loss: wear and tear in circulation, clipping and «sweating», and hoarding. Their chief protection against wear and tear was their high base-metal contents, less subject to such erosion than silver. But two countervailing factors offset such protection to some extent: (a) a relatively greater velocity of circulation (than for silver gros) in the payment of wages, in servicing local trade, particularly day to day transactions in foodstuffs; and (b) relatively more damage from purely chemical erosion (grease, dirt) upon the copper, a continuous process even when coins are stored, and a chemical process to which silver and gold were much more immune.[39] Nevertheless, on balance, copper or other base-metal coins evidently suffered less weight loss than more purely silver coins.[40]

The second major hypothesis, therefore, to explain why petty coinage accounted for such a small proportion of the Flemish mint outputs (or of the bullion struck) is that stocks of monnaies noires were much less subject to diminution from export or wear and tear losses than good silver coin, and consequently such stocks did not have to be replenished as frequently by the mints. This hypothesis probably has less «explanatory power» than the first; and neither hypothesis is put forward to the total exclusion of the other explanations discussed. This analysis suggests, therefore, that to a very large extent the production of petty coinage by the Flemish mints was in response to the market forces of supply and demand, even if the existing market structure, by so limiting direct public contact with the mints, transmitted general public demand for petty coinage very imperfectly indeed. Whether the Flemish monetary authorities, the counts, then Burgundian dukes truly met their public responsibilities in supplying sufficient petty coinage will certainly be questioned by many.

But in the mid-fifteenth century, in both England and the Low Countries, we find evidence that when the public was truly faced with a severe shortage of petty coinage the monetary authorities did attempt to meet those responsibilities. For both countries, indeed for northern Europe in general, the

[38] For this argument, see J.H. MUNRO, *Wool, Cloth, and Gold*, pp. 29-30 (see n. 5); R. DE ROOVER, *Gresham on Foreign Exchange*, Cambridge Mass., 1949, p. 82; A. FEAVEARYEAR, *Pound Sterling*, p. 22 (see n. 5).

[39] Ph. GRIERSON, *Coin Wear*, pp. VII-X (see n. 37).

[40] Ph. GRIERSON, Ibid., pp. X-XII.

middle decades, from the 1440s to the 1460s, were a period of severe monetary scarcity - a veritable «bullion famine,» according to some scholars - and of equally severe deflation.[41] Under such circumstances, the demand for petty coinage should have risen in accordance with its increased purchasing power, evident in Table 1. At the same time, obviously, a bullion famine would severely curtail the mints' ability to produce even petty coinage (which always contained at least some silver), particularly because the mints could remain open only so long as their operating costs were met by a sufficient influx of bullion and volume of coinage output. During these mid-century decades mint outputs were indeed very meagre across northern Europe, and many mints were forced to close.

In both England and the Burgundian Low Countries we find virtually the same positive response from the monetary authorities to such problems of coinage scarcity. During the 1440s, English silver coinage output, despite a relatively favourable mint ratio for that metal, had reached its lowest ebb since the beginning of the century (at the end of another and equally severe bullion famine): an annual average of 250.6 kg. of pure silver (= 1,068.6 marcs de Troyes AR). In the 1445 Parliament, a Commons petitioner complained about the «grete defaute» of silver amongst the poor people; and the Crown responded by decreeing a 10% increase in the number of halfpence and farthings (but not pence) struck from the Tower Pound of silver.[42]

The silver scarcity in Flanders during that same decade of the 1440s was even worse: an annual average of 119.8 kg. = 510.8 marcs AR minted, a drastic drop from the high mint outputs that had followed the Burgundian monetary reform of 1433-1434. In the ensuing decade, despite the plentiful coinages of the first half; only 1.9% of the total silver bullion at the mints had been coined into monnaies noires. But then in the 28 months from January 1445 to April 1447, when silver coinage temporarily ceased at the mints, a remarkable 70% of the silver bullion was minted into such petty coins, which thus accounted for 98.8% of the total number of coins struck, as Table 3

[41] See J. Day, *The Great Bullion Famine of the Fifteenth Century*, in: Past and Present 79 (1978), pp. 3-54, especially pp. 40-46; P. Spufford, *Monetary Problems*, pp. 118-121 (see n. 22); J.H. Munro, *Monetary Contraction*, pp. 109-110, 157 (Table 17) (see n. 19); Id., *Bullion Flows* (n. 17), pp. 119-126; Id., *Mint Outputs, Money, and Prices*, pp. 53-60 (see n. 12); Id., *Economic Depression*, pp. 235-250 (see n. 19).

[42] From February 1445 to October 1447, the London Tower Mint struck halfpence and farthings at 33s. instead of 30s. to the Tower Pound: thus increasing the number of halfpence from 720 to 792 per pound, and reducing the coin's weight by 9.09%, from 0.4860 g. to 0.4418 g. Note the reciprocal relationship between changes in silver content (X) and coin values: (1/1 - x) - 1. See *Rotuli Parliamentorum ut et petitiones et placita in parliamento*, VI, Londen, 1777, p. 108-109: n° 36; J.H. Munro, *Wool, Cloth, and Gold*, pp. 128-131 (see n. 5); Id., *Bullion Flows*, pp. 120-126, and Tables 1-3, pp. 131-135; Tables 7-8, pp. 142-146 (see n. 17); and the mint account sources cited in n. 17 above.

shows. In absolute amounts, however, that meant just 88 of the total 126 marcs AR struck into mites. In the next decade, as Table 3 also shows, a minor debasement of gold in January 1454 succeeded in reactivating the mints,[43] so that in just over four years, apart from the new gold coinage, some 3,503.4 marcs AR of silver (= 821.7 kg. pure) were minted, of which 1,455.1 marcs were struck into mites, or 41.5% of the total bullion - again, most remarkable in both absolute and relative terms (and representing 90% of the total number of coins struck). In part, a Burgundian monetary decree, similar to the English one, was responsible: an ordinance of 15 October 1456 requiring an 11.1% increase in the number of mites to be struck per marc AR (thus reducing their weight by 10%). For reasons not known to us, however, the Bruges mint effected this ordinance only on 3 September 1457, and then ceased coining not only mites but also all silver on 12 October 1458.[44] A trickle of gold coinage issued from the mint until March 1462, when all Burgundian mints were closed, reopening only with the general Burgundian debasement of June 1466.[45]

The Burgundian monetary ordinance of 23 May 1466, effecting the first general debasement of both silver and gold coinages since the Reform of October 1433, and Philip the Good's last such ordinance, is also the first to recognize a direct public interest in the coinage particularly in terms of its need for petty coinage. Perhaps to some considerable extent such concern reflects the fact that Duke Philip, responding to Edward IV's English debasements of August 1464 - March 1465, and to an influx of other foreign debased coins, had to seek agreement from the town assembles or «estates» for his own debasements. These assemblies, dominated by the leading Flemish towns, which had long claimed a share in the governance of Flanders itself, met seven times from January 1465 to May 1466 before the ordinance was finally promulgated. In such circumstances we have no grounds for dismissing its preamble as hypocrisy:[46]

> «que ung des principaulx poins de toutes bonnes pollicies sur quoy le bien publique, tant de nous comme du peuple, est fonde est de avoir et entretenir bonne monnoye, ferme et durable, tant dor comme dargent».

---

[43] J.H. MUNRO, *Wool, Cloth, and Gold*, pp. 149-150, 210 (see n. 5): by issuing a new gold coin, lion d'or, with a traite (→15.0. 0d.) that was 5.83% higher than that granted the previous gold coin, the Philippus (rijder), in 1443 (and 10.2% higher than its 1433 value).

[44] Ordinance cited in the Bruges mint account of 3 Sept. 1457 - 12 Oct. 1458: ARA, Rek. reg. n° 18, 105. See Tables 2-3; and see n. 42.

[45] J.H. MUNRO, *Wool, Cloth, and Gold*, pp. 155-158, 168-170; Tables A-K, pp. 187-211; ID., *Bullion Flows*, pp. 117-126 (see n. 17); Tables 3-10.

[46] ARA, Rek. reg. n° 133, f° 174^{V}; J.H. MUNRO, *Wool, Cloth, and Gold*, pp. 155-170, 178-179 (see n. 5).

The ordinance, after decreeing the finenesses, weights, and official values of the new coinages, gold and silver (from mites to double patards) then made some stipulations, the first such, about the composition of the silver coinage to ensure some supply of «small change»:[47]

> «affin que chacun puist estre pourvu de toutes manieres de deniers dor et dargent et de chacun diceulx, selon que la necessite le requiert, les maistres particuliers de la dite monnoie seront tenuz de forger»:

(1) for every marc of gold in florins, 100 marcs of silver in coin
(2) for every 100 marcs of silver patards (double gros):
- 10 marcs in double patards (of 4d. gros)
- 2 marcs in single gros
- 1 marc in demi-gros
- 1/2 marc in quarter-gros.

It would be difficult to imagine such an ordinance, such monetary requirements, in the reign of Count Louis de Male (1346–1384). One may also cavil, however, that it would be equally difficult to imagine these requirements under Maximilian of Habsburg, an equally rapacious manipulator of the mints. But a Burgundian-Habsburg monetary ordinance of April 1484, under the regency of the same Maximilian, contained the very same requirements: «so that the poor people may find themselves sufficiently supplied with *petite monnoie*», the mints were required to strike 50 marcs AR in single gros for every 100 marcs of patards (2d); and similarly, 10 marcs in demi-gros (12 mites), and 5 marcs in quarter-gros (6 mites).[48] Times had certainly changed, particularly with the increased political powers of the towns and their assemblies, which, from the later days of Duke Philip the Good (1419–1467), had never ceased to demand an equal role in determining monetary policies for the Low Countries.[49] As is so well known, of course, the

[47] ARA, Rek. reg. n° 133, f° 174^{v}–179^{v}; and also ARA, Rek. reg. n° 18.069, (Leuven mint account, 23 June 1466 - 7 July 1467); n° 18.196 (Ghent mint account, 18 June 1466–9 June 1467). See also L. Deschamps De Pas, *Essai sur l'histoire monétaire des comtes de Flandre de la maison de Bourgogne*, in: Revue Numismatique 2nd ser. 7 (1862), pp. 126–129.

[48] Text of the ordinance, in the name of Philippe le Beau (reproduced in L. Deschamps De Pas, *Essai sur l'histoire monétaire des comtes de Flandre de la maison d'autriche*, in: Revue Numismatique 2nd ser. 14 (1869–1870), p. 99). Deschamps de Pas is in error, however, in stating that this is the first such ordinance requiring a proportional striking of coins in this fashion. See also Ph. Grierson, *Numismatics*, p. 97; and for Maximilian's monetary policies and his relations with the town assemblies («estates»), see P. Spufford, *Monetary Problems*, pp. 141–146, 158–163 (see n. 22).

[49] See in particular P. Spufford, *Coinage, Taxation, and the Estates General of the Burgundian Netherlands*, in: Anciens Pays et Assemblées d'Etats (= Standen en Landen) 40 (1966), pp. 63–88; W. Blockmans, *La participation des sujets flamands à la politique monétaire des ducs de Bourgogne (1384–1500)*, in: Revue Belge de Numismatique (1973), pp. 103–134; W. Blockmans, *Autocratie ou polyarchie? La*

Flemish towns shortly thereafter revolted against Maximilian's rule, and peace was not restored until the 1490s. To what extent his sixteenth-century successors met their social responsibilities in matters of the coinage, especially in supplying sufficient petty coinage, will have to be determined by the scholars of this era. But now that all of the mint production and monetary statistics for the Low Countries have been collected, from 1334 to 1789, and fed into a computer for imminent publication, that task should not be so difficult.[50]

A Postscript on Gold Coinage

Gold coinage has been omitted from this discussion because its inclusion would introduce unnecessary and confusing complications. The principal tasks have been to determine the proportions of silver bullion and coinages struck in the form of petty coins, to plot changes in those proportions, and to suggest reasons for them. Those silver coins, from the basest mites to the purest silver patards, serviced the domestic needs of the medieval economy almost exclusively: the local market and retail trades, wage payments, etc. For most such purposes, gold coins were quite ill suited. Consider that in 1450 a single gold *Phillipus* (*rijder*), valued at 50d. gr., would have purchased 57 1b. of butter, 202.5 1b. of cheese, 38.5 gallons of wheat, 253.5 1b. of peas, 435 herrings, or 29 pts. of Rhine wine. While gold was therefore reserved generally for transacting wholesale and international trade, state finances and warfare, etc., silver could also be called upon occasionally to service such needs, when mint ratios altered strongly enough its favour, as noted earlier. Perhaps because of these reasons, silver's wider role in the Flemish economy, fluctuations in silver coinage outputs tended to be less sensitive (price-elastic) and gold coinages very much more sensitive to changes in the bimetallic ratio. The addition of the more volatile gold coinages would thus have greatly complicated our estimations of the proportions of the *total* value of the combined coinages accounted for by the petty coins; and obviously such proportions would have been infinitesimally - and unrealistically - small in times of high gold coinage outputs. There were two long periods in which Flemish mint ratios did favour gold strongly (in relation to other mints): ca. 1350–1384 and 1425–1465, during which gold coinage outputs were frequent-

---

lutte pour le pouvoir politique en Flandre de 1482 à 1492, d'après des documents inédits, in: Bulletin de la Commission Royale d'Histoire 140 (1974), pp. 257–368.

[50] To be published under the editorship of E. Van Cauwenberghe (Brussels-Leuven), R. Metz (Cologne), J. Munro (Toronto), F. Irsigler (Trier).

ly, but not always, high. Note once more, however, that that anomalous period of *relatively* high petty coinage outputs occurred in the middle of that second period, ca. 1445–1458. In sum, we are best advised to consider the Flemish petty coins within their own realm and family of the silver coinages.

SOURCES OF THE TABLES

Table 1: Commodity Prices

Ch. VERLINDEN, E. SCHOLLIERS, H. COPPEJANS-DESMEDT, J. CRAEYBECKX et al. (eds.), *Dokumenten voor de geschiedenis van prijzen en lonen in Vlaanderen en Brabant*, 4 vols., Bruges, 1965 [I, p. 44-45 (peas); II, p. 33-35 (wheat), p. 64-66 (butter), p. 67-69 (cheese)]
H. VAN DER WEE, *Growth of the Antwerp Market and the European Economy (Fourteenth to Sixteenth Centuries)*, 3 vols., The Hague, 1963 [I: Statistics, p. 277-281 (fresh herring), p. 294-298 (Rhine wine); p. 123-126 (monetary conversions)]

Tables 2-5: Mint Accounts of Flanders and Dependencies (1334-1384):

(1) Ghent, Bruges, and Mechelen, 1334-1384:
V. GAILLARD (ed.), *Recherches sur les monnaies des comtes de Flandre*, II: *Sous les règnes de Louis de Crécy et de Louis de Male*, Ghent, 1856; H. VAN WERVEKE, *De muntslag in Vlaanderen onder Lodewijk van Male* (Mededelingen van de Koninklijke Vlaamse Academie voor Wetenschappen, Letteren, en Schone Kunsten van België, Klasse der letteren, XI:5), Brussels, 1949.
F. and W.P. BLOCKMANS, *Devaluation, Coinage, and Seignorage under Louis de Nevers and Louis de Male, Counts of Flanders (1330-1384)*, in: Coinages in the Low Countries (880-1500). The Third Oxford Symposium on Coinage and Monetary History, (British Archeological Reports, International Series 54), Oxford, 1979, pp. 69-94.

(2) Ghent, Bruges, Mechelen, Valkenberg, Namur (13841484):
- a) Ghent: ARA [Algemeen Rijksarchief België, Archives Générales du Royaume], Rekenkamer, Rolrekening n° 824-826 (1388-1390); n° 827-831 (1410-1419); Acquits de Lille, liasses n° 936-937 (1410-1447); Rekenkamer, registers n° 580 (1419-1420), 18, 195-198 (1459-1462, 1466-1467, 1482-1484).
- b) Bruges: ARA, Rekenkamer, Rol. n° 776-787 (1392-1402); Acquits de Lille, liasse n° 1512 (1454-1456, 1466-1467); Rekenkamer, registers n° 18, 103-116 (1455-1482); Archives départementales du Nord (Lille), Chambre des Comptes, Série B, n° 644/15.939 (1454-1455), 19.960/19.312 (1454-1455).
- c) Mechelen: ARA, Rekenkamer, Rol. n° 2142-2143 (1384-1385), 2145-2146 (1390-1392); reg. n° 48,876-877 (1390-1392); Acquits de Lille, liasse n° 1512 (1454-1456, 1459-1460).

d) Valkenberg (Fauquemont): ARA, Rekenkamer, Rol. n° 2586-2587 (1396-1398).
e) Namur: ARA, Rekenkamer, reg. n° 18,203 (1421-1433).

Sources:

L. DESCHAMPS DE PAS, *Essai sur l'histoire monétaire des comtes de Flandre*, in: Revue Numismatique 2nd ser. 6 (1861), pp. 106-139, 211-237 and 458-478; ID. in: Revue Num. 2nd ser. 7 (1862), pp. 117-143, 351-365 and 460480; ID. in: Revue Num. 2nd ser. 11 (1866), pp. 177-219; ID. in: Revue Num. 2nd ser. 14 (1869-1870), pp. 86-113: texts of the monetary ordinances.

Table 6: Mint Ordinance of October 1467
ARA, Rekenkamer, register n° 580, f° 100; 18,106 (Bruges mint account: 23.01.1468-3.07.1469).
L. GILLIODTS-VAN SEVEREN (ed.), *Inventaire des archives de Bruges*, 6 vols. Bruges, 1871-1878, [V, doc. n° 1101, pp. 547-554].

Table 1
The Purchasing Power of One Flemish Silver Gros (1d) in Terms of Selected Foodstuffs in the Low Countries : 20 Year Intervals from 1350 to 1485

YEAR	BUTTER (Bruges) in kg./lb.	CHEESE (Bruges) in kg./lb.	WHEAT[a] (Bruges) litres/gallons	WHEAT[a] (Bruges) kg./lb.	PEAS[a] (Bruges) litres/gallons	PEAS[a] (Bruges) kg./lb.	HERRING FRESH (Antwerp) number	RHINE WINE (Ghent) litres pints
1350	0.93 kg. (2.05 lb.)	2.33 kg. (5.14 lb.)	4.77 l. (1.05 g.)	3.57 kg. (7.88 lb.)				
1370	0.40 kg. (0.88 lb.)	0.99 kg. (2.19 lb.)	2.05 l. (0.45 g.)	1.52 kg. (3.54 lb.)				
1390	0.36 kg. (0.79 lb.)	0.98 kg. (2.15 lb.)	2.46 l. (0.54 g.)	1.84 kg. (4.05 lb.)			7.2	
1410	0.49 kg. (1.07 lb.)	1.24 kg. (2.73 lb.)	2.41 l. (0.53 g.)	1.81 kg. (4.00 lb.)	3.59 l. (0.79 g.)	2.68 kg. (5.91 lb.)	8.3	0.46 l. (0.81 pt.)
1430	0.29 kg. (0.63 lb.)	1.06 kg. (2.34 lb.)	2.95 l. (0.65 g.)	2.22 kg. (4.89 lb.)	2.95 l. (0.65 g.)	2.22 kg. (4.89 lb.)	6.3	0.27 l. (0.48 pt.)
1450	0.43 kg. (0.94 lb.)	1.84 kg. (4.05 lb.)	3.50 l. (0.77 g.)	2.63 kg. (5.79 lb.)	3.09 l. (0.68 g.)	2.30 kg. (5.07 lb.)	8.7	0.33 l. (0.58 pt.)
1470	0.39 kg. (0.85 lb.)	2.25 kg. (4.95 lb.)	3.73 l. (0.82 g.)	2.80 kg. (6.17 lb.)	3.59 l. (0.79 g.)	2.68 kg. (5.91 lb.)	5.9	0.31 l. (0.54 pt.)
1485	0.36 kg. (0.79 lb.)	1.32 kg. (2.90 lb.)	3.45 l. (0.76 g.)	2.58 kg. (5.68 lb.)	2.95 l. (0.65 g.)	2.22 kg. (4.89 lb.)	5.0	0.21 l. (0.37 pt.)

[a] 1 Imperial gallon = 4.546 litres. 8 gallons = 1 imperial bushel = 60 lb. wheat or peas = 27.2 kg.

Table 2
The Coinage of Double Mites or Courtes in the Flemish Mints, 1334-1484:
Weight (Taille per Marc), Fineness, Silver Content of the Coin, Seignorage, and Brassage

Dates Struck	Taille to the Marc	Percentage Fineness Argent-le-Roy	Grams of Pure Silver in the Coin	Traite of the Marc Argent-le-Roy in s.	Seignorage per Marc Argent-le-Roy in d.	Brassage per Marc Argent-le Roy in s.	Total Mintage as % of the Traite
01.10.1334-07.11.1338	264	14.583%	0.130g.	6.286s[a]	nil	n.a.	n.a.
12.01.1375-06.11.1376	250	6.250%	0.059g.	27.778s.	12d.	n.a.	n.a.
06.11.1376-27.06.1377	250	6.250%	0.059g.	27.778s.	8d.	n.a.	n.a.
01.02.1389-10.12.1389	240	4.167%	0.041g.	40.000s.	12d.	17s.0d.	45.00%
03.11.1390-20.06.1393	183	4.167%	0.053g.	30.500s.	2d.	13s.4d.	44.26%
20.06.1393-16.06.1402	183	4.167%	0.053g.	30.500s.	2d.	12s.10d.	42.62%
22.11.1419-09.11.1428	180	3.125%	0.041g.	40.000s.	4d.	18s.6d.	47.08%
09.03.1430-27.09.1432	180	2.777%	0.036g.	45.000s.	7d.	20s.5d.	46.67%
15.10.1433-26.07.1457	216	4.167%	0.045g.	36.000s.	2d.	13s.1d.	36.81%

Table 2, continued

Dates Struck	Taille to the Marc	Percentage Fineness Argent-le-Roy	Grams of Pure Silver in the Coin	Traite of the Marc Argent-le-Roy in s.	Seignorage per Marc Argent-le-Roy in d.	Brassage per Marc Argent-le Roy in s.	Total Mintage as % of the Traite
03.09.1457-12.10.1458	240	4.167%	0.0407g.	40.000s.	2d.	17s.1d.	43.13%
18.06.1466-09.06.1467	198	3.472%	0.0411g.	39.600s.	1½d.	13s.1d.17m.	33.50%
23.01.1468-10.04.1474	204	3.472%	0.0400g.	40.800s.	1½d.	13s.9d.2m.	34.03%
20.12.1477-05.04.1482	216	2.777%	0.0300g.	54.000s.	6d.	23s.2d.	43.83%
18.07.1482-21.05.1484	196	2.083%	0.0250g.	65.333s.	3¼d.	26s.8d.18m.	41.33%

a Probably struck as single rather than double mites. If double mites: a traite of 12.571s. gros. One courte or double mite = 2 mites = 1d. parisis = 1/12d. gros Flemish.

Table 3
The Petty Silver Coinage of Late-Medieval Flanders
The Coinage of Mites, Double Mites, Quarter Gros and of Total Silver in Flemish Mints: in Marcs Argent-le-Roy, in Values of Output (Pounds Gros Flemish), and Number of Coins Struck, in Ten-Year Periods, from 1334 to 1484

PERIOD Ten Years (approx.)	IN MARCS ARGENT-LE-ROY[a]			IN POUNDS GROS FLEMISH			IN NUMBER OF COINS STRUCK		
	Petty Coinage	Total Coinage (Silver)	Petty as % of Total	Petty Coinage	Total Coinage (Silver)	Petty as % of Total	Petty Coinage	Total Coinage (Silver)	Petty as % of Total
16.09.1334-17.10.1343	1,350.0	116,295.4	1.16%	424.3	35,406.7	1.20%	2,443,848	15,744,031	15.52%
20.01.1346-24.10.1354	Nil	203,697.9	0	Nil	98,645.0	0	Nil	24,796,048	0
20.12.1354-02.03.1364	Nil	254,534.0	0	Nil	144,877.2	0	Nil	34,687,263	0
12.04.1365-26.06.1374	Nil	277,457.3	0	Nil	209,977.0	0	Nil	31,652,241	0
26.06.1374-29.02.1384	408.3	97,463.8	0.42%	567.2	91,576.2	0.62%	1,633,250	12,554,345	13.01%
29.02.1384-24.06.1394	347.8	102,096.3	0.34%	548.5	99,229.3	0.55%	1,963,238	16,267,260	12.07%
24.06.1394-16.06.1402	693.9	161,946.8	0.43%	1,056.8	153,883.0	0.69%	4,454,090	23,681,485	18.81%
17.01.1411-[b] 29.04.1415	791.7[c]	76,540.7	1.03%	682.8[c]	65,987.9	1.03%	655,500[c]	9,098,274	7.20%
30.10.1417-[d] 10.11.1424	2,117.5[e] [260.2][f]	285,153.1	0.74% [0.09%]	2,859.3[e] [537.8][f]	323,583.9	0.88% [0.17%]	3,777,456[e] [1,548,756][f]	44,564,218	8.48% [3.48%]

Table 3, continued

PERIOD Ten Years (approx.)	IN MARCS ARGENT-LE-ROY[a]			IN POUNDS GROS FLEMISH			IN NUMBER OF COINS STRUCK		
	Petty Coinage	Total Coinage (Silver)	Petty as % of Total	Petty Coinage	Total Coinage (Silver)	Petty as % of Total	Petty Coinage	Total Coinage (Silver)	Petty as % of Total
10.11.1424-[d] 14.10.1434	403.0[e] [165.2][f]	247,036.2	0.16% [0.07%]	657.9[e] [344.8][f]	305,267.1	0.22% [0.11%]	1,293,630[e] [993,042][f]	39,110,243	3.31% [2.54%]
15.10.1434- 31.12.1444	1,966.9 [161.2][f]	105,293.7	1.87% [0.15%]	2,641.7 [285.3]	126,639.5	2.09% [0.23%]	3,182,730 [920,556][f]	18,443,598	17.26% [4.99%]
29.01.1445- 05.04.1447[g]	88.0[f]	126.0	69.83%	158.3[f]	203.9	77.64%	455,976[f]	461,448	98.81%
02.03.1454- 16.03.1462	1,455.1[e] [399.0][f]	3,503.4	41.53% [11.39%]	2,258.6 [880.4][f]	4,752.2	47.53% [18.53%]	4,837,881 [3,514,872][f]	5,406,317	89.49% [65.01%]
18.06.1466- 10.04.1474	3,282.8[e] [449.8][f]	153,298.6	2.14% [0.29%]	5,212.0 [909.3][f]	217,295.9	2.40% [0.42%]	6,749,201 [2,618,565]	29,506,127	22.87% [8.87%]
20.12.1474- 21.05.1484	712.4[e] 458.2][f]	132,655.1	0.54% [0.35%]	1,939.7 [1,386.7][f]	230,342.1	0.84% [0.60%]	4,621,935 [3,993,685]	30,089,292	15.36% [13.27%]

[a] 1 Marc de Troyes = 244.753 grams. Argent-le-Roy = 23/24ths or 95.833% fine.
[b] Flemish mints closed from 17.06.1402 to 16.01.1411.
[c] Quarter-gros coins only struck.
[d] Includes output of the Namur mint.
[e] Includes mites and quarter-gros.
[f] Mites and Double Mites only (i.e. excluding quarter-gros).
[g] No silver coinage from 05.04.1447 to 02.03.1454.

Flemish and auxiliary mints included:

Ghent : 16.09.1334 - 27.06.1377; 01.10.1388 - 23.10.1390; 17.01.1410 - 05.04.1447; 02.05.1459 - 16.03.1462; 18.06.1466 - 09.06.1467; 18.07.1482 - 21.05.1484
Bruges : 20.08.1350 - 24.10.1354; 01.06.1392 - 16.06.1402; 02.03.1454 - 05.04.1482
Mechelen : 08.10.1357 - 02.03.1364; 30.01.1380 - 29.10.1392; 16.06.1454 - 16.01.1460
Falkenberg: 01.11.1396 - 01.11.1399
Namur : 23.10.1421 - 10.10.1433

Table 4
The Coinage of Quarter Gros (Gigots or Zeskins) in the Flemish Mints, 1334-1484: Weight (Taille per Marc), Fineness, Silver Content of the Coin, Seignorage, and Brassage

Dates Struck	Taille to the Marc	Percentage Fineness Argent-le-Roy	Grams of Pure Silver in the Coin	Traite of the Marc Argent-le-Roy in s.	Seignorage per Marc Argent-le-Roy in d.	Brassage per Marc Argent-le Roy in s.	Total Mintage a % of the Traite
01.10.1334-08.06.1336	230	85.417%	0.871g.	5.610s.	1d.	?	?
17.01.1410-29.04.1415	276	33.333%	0.283g.	17.250s.	2d.	1s.11d.	12.08%
14.08.1420 09.11.1428	300	25.000%	0.195g.	25.000s.	4d.	3s.6d.	15.33%
09.03.1430-27.09.1432	288	22.222%	0.181g.	27.000s.	7d.	2s.5d.	11.11%
25.10.1433-30.06.1456	348	27.777%	0.187g.	26.100s.	2d.	3s.2d.5m.	12.84%
18.06.1466[a] 09.06.1467	348	25.000%	0.169g.	29.000s.	1½d.	2s.6d.12m.	9.20%
04.07.1469-14.04.1472	324	22.222%	0.161g.	30.375s.	1½d.	3s.4d.	11.38%
20.12.1477-05.04.1482	310	16.667%	0.126g.	38.750s.	6d.	7s.11d.	21.72%
18.07.1482[a] 12.06.1485	306	14.583%	0.112g.	43.714s.	3¼d.	5s.1d.8m.	12.31%

[a] The coinage of quarter-gros was officially prescribed at the stated fineness and taille; but the mint accounts for the periods given state that none was struck.
Quarter-gros (gigot or zeskin) = 6 mites = 3d. parisis = 1/4d. gros of Flanders.

Table 5
A Comparison of the Petty and Gros Silver Coinages Struck in the Flemish Mints, 1334-1484: Traites per Marc, Silver Contents, and Seignorage

PERIOD OF MINTING	PETTY COINAGES (Mites, Quarter-Gros)				GROS COINAGES (Single, Double)			
	Petty Coin	Traite of Petty coin per Marc Argent-le-Roy in s.	Seignor-age per Marc Argent-le-Roy	Silver Content of the Petty Coin in g.	Silver Content as % of Silver in the Gros	Silver[a] Content of the Gros in g. (1d)	Traite of Gros per Marc Argent-le-Roy in s.	Seignor. per Marc Argent-le-Roy
01.10.1334-25.05.1337	M[b] QG	6.286s. 5.610s.	Nil 1d.	0.130g. 0.871g.	3.73% 25.00%	3.484g.	5.610s.	1d.
25.05.1337-07.11.1338	M	6.286s.	Nil	0.130g.	4.47%	2.908g.	6.722s.	7d.
12.01.1375-06.11.1376	DM	27.778s.	12d.	0.059g.	5.30%	1.114g.	17.538s.	2d.
06.11.1376-27.06.1377	DM	27.778s.	8d.	0.059g.	5.30%	1.114g.	17.538s.	2d.
01.02.1389-10.12.1389	DM	40.000s.	12d.	0.041g.	5.25%	0.781g.	25.034s.	6d.
03.11.1390-16.06.1402	DM	30.500s.	2d.	0.053g.	5.15%	1.029g.	19.000s.	4d.
17.01.1410-29.04.1415	QG	17.250s.	2d.	0.283g.	23.64%	1.197g.	16.333s.	2d.
22.11.1419-09.11.1428	DM QG	40.000s. 25.000s.	4d. 4d.	0.041g. 0.195g.	4.76% 22.62%	0.862g.	22.667s.	4d.
09.03.1430-27.09.1432	DM QG	45.000s. 27.000s.	7d. 7d.	0.036g. 0.181g.	4.73% 23.78%	0.761g.	25.6875s.	6d.

Table 5, continued

PERIOD OF MINTING	PETTY COINAGES (Mites, Quarter-Gros)				GROS COINAGES (Single, Double			
	Petty Coin	Traite of Petty coin per Marc Argent-le-Roy in s.	Seignor-age per Marc Argent-le-Roy	Silver Content of the Petty Coin in g.	Silver Content as % of Silver in the Gros	Silver[a] Content of the Gros in g. (1d)	Traite of Gros per Marc Argent-le-Roy in s.	Seignor. per Marc Argent-le-Roy
15.10.1433- 30.06.1457	DM QG	36.000s. 26.100s.	2d. 2d.	0.045g. 0.187g.	5.53% 22.97%	0.814g.	24.000s.	2d.
03.09.1457- 12.10.1458	DM	40.000s.	2d.	0.0407g.	5.00%	0.814g.	24.000s.	2d.
18.06.1466- 09.06.1467	DM QG[c]	39.600s. 29.000s.	1½d. 1½d.	0.0411g. 0.169g.	5.78% 23.77%	0.711g.	27.500s.	1½d.
23.01.1468- 10.04.1474	DM QG	40.800s. 30.375d.	1½d. 1½d.	0.040g. 0.161g.	5.76% 23.20%	0.694g.	28.167s.	1½d.
20.12.1474- 20.12.1477	d					0.611g.	32.000s.	6d.
20.12.1477- 05.04.1482	DM QG	54.000s. 38.750s.	6d. 6d.	0.030g. 0.126g.	4.91% 20.62%	0.611g.	32.000s.	6d.
18.07.1482- 21.05.1484	DM QG[c]	65.333s. 43.714s.	3¼d. 3¼d.	0.025g. 0.112g.	5.11% 22.90%	0.489g.[e]	40.000s.	3¼d.

Table 5, continued

[a] The silver contents of the Flemish gros (<u>groot</u>) coinages are reckoned in terms of the single gros up to the monetary reform of 1390 (except for 1375-1377, when only double gros <u>lions</u> were struck), and thus represent the actual silver content of the gros. From the 1390 reform, silver contents are reckoned in terms of the standard of the double gros (<u>patard</u>), and represent the silver content in 1d gros (money-of-account), not in the actual gros coin. The silver contents are expressed in terms of pure silver, not argent-le-roy (which is 23/24ths or 95.833% pure silver).

[b] These coins were probably single mites; but if double mites, then a traite of 12.571s. gros.

[c] Although the mint indentures prescribed the striking of quarter-gros as specified in the table, the mint accounts state that no quarter-gros were struck for the period stated.

[d] No quarter-gros or mites were prescribed by the mint indentures, and none was struck in this period.

[e] The silver content of the <u>patard</u> was not altered, but its rate was raised from 2d. to 2½d.

Table 6
The Burgundian Monetary Ordinance of 13 October 1467[a]

A. Silver DOUBLE PATARD: worth 4d. gros
Fineness: 11 deniers argent-le-roy: 11/12 or 91.67% fine = 87.85% pure
Weight: 77 1/2 struck to the marc de Troyes = 3.158 g.
Pure Silver Content: 2.774 g.

Division of the Traite	Value in Flemish Gros	No. of Coins	Percentage of Traite
Brassage	8d. 12 mites	2 1/8	2.52%
Seignorage	1d. 12 mites	3/8	0.44%
Total Mint Charges	10d.	2 1/2	2.96%
Mint Bullion Price	27s. 4d.	82	97.04%
TRAITE	28s. 2d.	84 1/2	100.00%

B. DOUBLE MITE or COURTE: worth 1/12th d. gros = 1d. parisis
Fineness: 10 grains argent-le-roy: (10/24)/12 = 3.47% fine = 3.33% pure
Weight: 204 struck to the marc de Troyes = 1.200 g.
Pure Silver Content: 0.040 g.

Division of the Traite	Value in Flemish Gros	No. of Coins	Percentage of Traite
Brassage	13s. 9d. 2 mites	1,981	33.72%
Seignorage	1d. 12 mites	18	0.31%
Total Mint Charges	13s. 10d. 14 mites	1,999	34.03%
Mint Bullion Price	26s. 11d.	3,876	65.97%
TRAITE	40s. 9d. 14 mites	5,875	100.00%

[a] Period of Minting: 23 January 1468 to 10 April 1474

BULLION TRANSFER FROM WESTERN EUROPE TO THE BALTIC AND THE PROBLEM OF TRADE BALANCES: 1550–1750

by Michael NORTH

During the sixteenth century the Baltic region became the predominant supplier of grain on the rising European world market. Moreover the societies of the Baltic satisfied Western Europe's growing demand for industrial raw materials such as flax and hemp and for timber and such by-products as pitch, tar and potash. In exchange Western Europe supplied the Baltic with salt, herring, cloth, wine, fruits, spices and other colonial commodities. The striking feature of the trade with the Baltic was that the value of the Baltic exports exceeded by far the value of the imports from Western Europe. That is why Western Europe had to settle her negative balance in the Baltic trade to a great extent by payments in precious metals.[1] From the second half of the sixteenth century the Dutch played the dominant role in bullion trade and transport not only to the Polish and Russian markets but handled also the precious metals supply in the trade with the Levant and the East Indies. The purpose of our paper is first to quantify the volume of bullion flowing into the Baltic countries and second to trace the flow of precious metals in the coin hoards of Poland, thus examining the mechanisms of influx and trying to outline the external and internal factors which determined the influx of precious metals in quantity and quality.

Let us first touch the problem of the registration or better the non-registration of precious metals in the customs accounts.[2] In Europe neither coins nor unminted precious metals were regarded as merchandise and were therefore not subject to the Sound and other tolls. That is why the imports of precious metals – with the exception of the Narva tolls 1603–1608, 1611 and of the customs accounts along the border to Moldavia[3] – did not enter into

[1] A. ATTMAN, *The Bullion Flow between Europe and the East (1000–1750)*, (Acta Regiae Societatis Scientiarium et Litterarum Gothoburgensis. Humaniora 20), Gothenburg, 1981. A. ATTMAN, *Dutch Enterprise in the World Bullion Trade (1550–1800)*, (Acta Regiae Societatis Scientiarum et Litterarum Gothoburgensis. Humaniora 23), Gothenburg, 1983.

[2] A. ATTMAN, *Dutch Enterprise*, pp. 22 ff.

[3] Z. ŚWITALSKI, *Cło od pieniędzy wywożonych za granicę Rzeczypospolitej w latach (1598–1659)*, in: Przegląd Historyczny 51 (1960), pp. 24 ff.

toll registers thus depriving us of a valuable source. However there are contemporary estimates and present calculations which substitute for the non-existing information of the toll accounts to a considerable extent.

Very valuable is a memorandum on the export of precious metals submitted to the States-General by Dutch mint masters in 1683.[4] Therein the mint masters calculate an annual bullion import from Spain of 15 to 18 million guilders and an export of precious metals as shown in table 1.

Table 1

The East Indies	ca. 2m. guilders	(0.8m. rix-dollars)
Russia	ca. 2m. guilders	(0.8m. rix-dollars)
Poland, Danzig Königsberg, Memel etc.	ca. 1m. guilders	(0.4m. rix-dollars)
Germany	1m. guilders	(0.4m. rix-dollars)
Smyrna, Levant	2m. guilders	(0.8m. rix-dollars)
France	3m. guilders	(1.2m. rix-dollars)
England	2m. guilders	(0.8m. rix-dollars)
Total	13m. guilders	(5.2m. rix-dollars)

However the exclusion of Riga distorts the picture of bullion exports to the Baltic markets, since a Baltic export surplus of about 1 million rix-dollars is omitted and has to be added to the figures of the Baltic region.[5]

Most useful are the estimates by Attman, as they reflect the development of Dutch bullion exports to the Baltic area (including Archangel) over a period of nearly 200 years.[6]

Unfortunately we have not enough evidence of the distribution of bullion influx in the different Baltic trade areas. But we can state that during the eighteenth century, due to the expansion of the English-Russian trade, Riga and St. Petersburg realized the biggest export surpluses of the Baltic region, thus consuming most of the bullion shipped to the Baltic for the settlement of the balances. The deficits in England's Russian trade were counterbalanced by Dutch coins or by bills of exchange drawn on Amsterdam, bills that could be used as cash in the Baltic ports and were payable in Amsterdam from where the money was sent to the Baltic.[7]

---

[4] J.G. VAN DILLEN (ed.), *Bronnen tot de geschiedenis der wisselbanken*, I, 's Gravenhage, 1925, pp. 216 ff; A. ATTMAN, *Dutch Enterprise*, pp. 37 ff.
[5] A. ATTMAN, *Dutch Enterprise*, p. 40.
[6] Ibid., p. 91.
[7] Ibid., pp. 45 ff.

Table 2
Annual supplies of Dutch precious metals to the Baltic area (in millions of rix-dollars)

Around 1600	2
Around 1640	2.5
Around 1700	2
Around 1750	2
Around 1780	3

Now we will follow up the traces of Dutch precious metals in the coin hoards of Poland, giving first a brief outline of the changing - with respect to quality - influx of Dutch coins.[8] From the 1590s we notice a growing influx of Dutch dollars (thaler) and ducats into Poland. Among those, dollars of the Southern Netherlands dominated (Filipsdaalder and Bourgondische rijksdaalder of the Brabant and Geldern mints). But already in the last days of the sixteenth century rijksdaalders of the Northern Netherlands became more important, emerging as the prevalent heavy silver coins in the first half of the seventeenth century on the Polish money market. However the picture changed once more in the second half of the seventeenth century, when the dollars of the Spanish Netherlands, especially the patacons, the so-called Albertus- or Crossdollars, dominated again the heavy-silver-coin circulation in Poland. As regards the Dutch ducats we witness ducats exclusively of the Northern Provinces penetrating into Poland during the whole seventeenth and eighteenth centuries.

At the beginning of the seventeenth century Dutch ducats succeeded in displacing the Hungarian ducats which had been prevalent in Polish circulation until that time.[9] Moreover we notice exceptional phases of gold, i.e. ducat, influx in the middle of the seventeenth century and during the eighteenth century as well, when the flow of dollars drained off.

---

[8] A. Mikołajczyk, *Obieg pieniężny w Polsce środkowej w wiekach od XVI do XVIII*, (Acta Archaeologica Lodziensia 28), Łódź, 1980; M. Męclewska and A. Mikołajczyk, *Skarby monet z lat 1500-1649 na obszarze PRL*, Inventarz, Warszawa, 1983; A. Mikołajczyk, *W kwestii znalezisk nowożytnych monet niderlandzkich na ziemiach koronnych*, in: Zapiski Historyczne 49 (1984), n° 2, pp. 5 ff. For coin hoards and monetary circulation in Ducal Prussia see M. North, *Geldumlauf und Wirtschaftsregion. Untersuchungen am Beispiel Königlich Preußens und des Herzogtums Preußen in der Frühen Neuzeit*, in: Hamburger Beiträge zur Numismatik 30/32 (1976–1978, 1985), pp. 71 ff.

[9] A. Mikołajczyk, *Złote monety w nowożytnych znaleziskach ziem koronnych: próba rekonstrukcji ich obiegu*, in: Wiadomości Numizmatyczne 22 (1978), p. 71

The composition of the coin hoards reflects on the one hand monetary policy and monetary circulation in the Low Countries and on the other hand the variations of the Polish money market. Dutch monetary policy maintained real influence on the monetary policy in Poland, whilst Dutch merchants settling the trade balance deficits tried to adapt themselves to the demands of the Polish money market. The crucial point in the monetary policy of the Low Countries was the monetary reform of 1612 in the Southern Netherlands.[10] Two new heavy silver coins, the patacon and the ducaton, were created at rates 4% lower than the heavy silver coins of the Northern Provinces. The patacons and ducatons were very successful in the Southern Netherlands and penetrated as well into the Northern Netherlands becoming there the dominant heavy silver coins. In consequence the better rixdollars of the Northern Provinces, which were now undervalued, were taken out of circulation by Dutch merchants and exported as «negotiepenningen» into the Baltic countries which accepted them for their real value. However the Northern Provinces created by the monetary ordinance of 1659 two new silver coins, the rijder and the ducaat, imitating the Southern patacon and ducaton and pushing the patacons and ducatons out of circulation, since the new coins were overvalued compared with the patacons and ducatons. Thus from the middle of the seventeenth century patacons became more and more used in the settlement of Western Europe's balances in the Baltic trade.

Analyzing the variations of the Polish money market, we can state that the exchange rates of silver and gold in general exceeded the rates in Western Europe, whereas the gold-silver ratio underwent significant fluctuations, favouring at one time the import of silver and another time the import of gold. In the 1630s the Polish exchange rate of the golden ducat was 120% higher than in Germany and 37% higher than in the Low Countries and France; the Polish thaler rate of exchange exceeded the rates in Germany and in the Low Countries by 100% and by 25% respectively.[11] It is therefore not surprising that Dutch merchants made good use of these favourable rates, paying their grain and timber bills with ducats, which were exported in enormous quantities to Poland. This trend was interrupted during the second half of the seventeenth century when due to the rising gold prices in Western Europe the Dutch preferred silver for their payments. Again at the end of the century and throughout the eighteenth century the Polish money market attracted Dutch ducat imports which resulted in the dominance of Dutch ducats in Polish monetary circulation.

[10] For the following see H. VAN DER WEE, *Money and Economic Interdependance Between the Northern and Southern Netherlands and the Baltic (15th–17th Centuries)*, in: The Interactions of Amsterdam and Antwerp with the Baltic Region (1400–1800), Leiden, 1983, pp. 11 ff.

[11] A. MIKOŁAJCZYK, *W kwestii*, pp. 27 f.

After this illustration of Dutch monetary policy and Polish exchange rate development let us examine the geographical distribution of Dutch precious metals in the hoards of the Polish-Lithunian territory. During the seventeenth century we notice a significant shift in the concentration of gold and heavy silver coins from the North to the South-East of Poland. In the late sixteenth and early seventeenth century Dutch thalers and ducats were nearly exclusively treasured in Royal Prussia, i.e. the area near to Danzig and close to the Baltic Sea, whilst the 1630s and especially the 1640s saw a strong influx of gold and silver into the south-eastern parts of Poland, into Wolhynia, Podolia and the Ukraine; a trend which continued throughout the century, although the bullion imports into Poland decreased in the second half. From the 1630s precious metals were also channeled into Lithunia, but to a smaller extent in comparison with the Polish South-East. A particular position took Ducal Prussia that participated only a little of the bullion imports, even though the Duchy was included in the Polish-Lithunian market zone and the exchange rates of the Ducal Prussian money market roughly corresponded to the Polish exchange rates.[12]

In explaining the regionally different amounts of precious metals flowing into Poland and Ducal Prussia, we have to focus our attention on world-market production and export surplus realization by the regional economies. For example in the late sixteenth and early seventeenth century the overwhelming part of Poland's grain exports was cultivated in Royal Prussia which thus absorbed most of the bullion imported into Poland.[13] But in the following time the south-eastern areas became more and more integrated into the international division of labour, due to the expansion of grain production and cattle breeding (oxen), whereas Royal Prussia's grain supply for the world market diminished. Thus precious metals were channeled into Wolhynia, Podolia and the Ukraine, which accumulated the imported bullion only if they had an active export surplus. As regards the very low share of gold and heavy silver coins in Ducal Prussian hoards it resulted from the small extent of Ducal Prussia's integration into the world market, as a considerable part of the territory's grain production was based on the inland market's demand.[14] Whether Ducal Prussia had an export surplus, is not easy

---

[12] G. Meinhardt, *Die Münz- und Geldgeschichte des Herzogtums Preußen (1569–1701)*, (Studien zur Geschichte Preußens 4), Heidelberg, 1959, pp. 158 f., (Table IVa).

[13] A. Wyczański, *Próba oszacowania obrotu żytem w Polsce XVI wieku*, in: Kwartalnik Historii Kultury Materialnej 9 (1961), pp. 29 ff.

[14] M. North, *Die Amtswirtschaften von Osterode und Soldau. Vergleichende Untersuchungen zur Wirtschaft im frühmodernen Staat am Beispiel des Herzogtums Preußen in der zweiten Hälfte des 16. und der ersten Hälfte des 17. Jahrhunderts*, (Gießener Abhandlungen zur Agrar- und Wirschaftsforschung des europäischen Ostens 118), Berlin, 1982, pp. 34 ff.; --- , *Untersuchungen zur adligen Gutswirtschaft im Herzogtum Preußen des 16. Jahrhunderts*, Vierteljahresschrift für Sozial- und Wirtschaftsgeschichte 70 (1983), pp. 16 ff.

to judge, since the Baltic countries' trade and as well their balances of trade were multilateral.

However in the case of Ducal Prussia and of most of the Baltic countries we have only evidence of the countries' sea-borne trade, i.e. the trade balance of the ports. The differences between the balances of the ports and the total balances of the countries can be illustrated by the examples of the Ducal Prussian port Königsberg and of the Polish-Lithuanian balance of trade. By the middle of the seventeenth century Königsberg had an obvious export surplus, as the value of the exports exceeded that of the imports by 30% to 40%.[15] But the precious metals imported into Königsberg were needed for the settlement of Königsberg's deficits in her Lithuanian trade, since Königsberg's export surpluses were above all produced in the Lithuanian hinterland. Lithunia for her part sent the money eastwards to settle the negative balance in the Polish-Lithunian trade with Russia. In the eighteenth century too the imports of precious metals into Königsberg - according to J.J. Oddy worth of 2 million rixdollars[16] - were not treasured in Prussia but went in transit to Russia, as Königsberg played a crucial role in the transfer of bullion to the Russian market.

Striking are also the changes within Poland's multilateral trade balance. During the period 1550–1650 Poland's Baltic exports via Danzig, similar to the other Baltic ports, were about twice as great as the imports in term of value.[17] The trade balance with Germany seems to have been positive, whereas the deficite trade with Russia required precious metals to flow to the East. Poland's balance in her trade with Turkey was also negative, as enormous exports of money to Turkey are registered in the accounts of the customs stations along the Polish-Moldavian border.[18] In the middle of the seventeenth century Poland lost her export surplus, since Danzig's sea-borne balance became negative. Moreover the growing imports across the land borders, from the West and from the Orient, contributed to the negative total balance of the Polish Republic of Nobles. This meant a slackening influx of precious metals, of which the composition of the coin hoards gives clear evidence.[19] Striking are the long periods between the issue of the late seventeenth century dollars and their deposit in the earth (see Table 1) which can not be explained by the dollars' long circulation, since the dollars do not

---

[15] H. Kempas, *Seeverkehr und Pfundzoll im Herzogtum Preußen. Ein Beitrag zur Geschichte des Seehandels im 16. un 17. Jahrhundert*, Diss. phil. Bonn, 1964, pp. 255 ff.

[16] J.J. Oddy, *European Commerce*, London, 1805, p. 232.

[17] A. Attman, *The Bullion Flow*, p. 86; A. Mączak, *Między Gdanskiem a Sundem. Studia nad handlem bałtyckim od połowy XVI do połowy XVII wieku*, Warszawa, 1972, pp. 69 ff.

[18] Z. Switalski, *Cło od pieniędzy*, pp. 27 f.

[19] A. Mikołajczyk, *W kwestii*, pp. 23 f.

show the characteristic signs of permanent circulation. However the long periods between issue and deposit indicate that the thalers were collected in the Low Countries and sent only to Poland if they were required, as the shipments of precious metals to Poland had ceased being regular.

Table 3
Periods of issue and thesauration of Dutch dollars hoarded in Poland (in per cent)

period	issue	thesauration
1551-1575	5.3	-
1576-1600	6.2	7.3
1601-1625	28.5	1.4
1626-1650	33	14.4
1651-1675	17.3	16.2
1676-1700	9.7	33.5
1701-1725	-	19.2
1726-1750	-	6.6
1751-1775	-	0.9
1776-1800	-	0.3
1801-1825	-	0.2
summa	100.0	100.0

Source: A. Mikołajczyk, W kwestii, p. 24 (table 7)

Summarizing our inquiries we can state that a country's trade balance with the West determined the bullion influx in quantity, whereas the country's total balance decided to what extent the precious metals remained in the country. Whether gold or heavy silver coins - and which heavy silver coins - were consigned to the Baltic depended first of all on the demands of the Polish money market - manifest to the merchants in the exchange rates and the gold-silver ratios - and secondly was influenced by monetary policy and mint output in the Low Countries. However, whether a region profited by the influx of precious metals depended on the extent of integration of the region's productive capacities into the world market, its participation in the international division of labour.

THE BULLION FLOW FROM EUROPE TO THE EAST: 1500–1800

by Artur ATTMAN

During the early Middle Ages a balance problem arose in European world trade: it persisted for several centuries, and related to deficits in the Levantine trade and the trade with Russia. In the Italian cities' trade with the Orient and in Lübeck's trade with Novgorod and Riga, the deficits were settled with supplies of precious metals (especially silver) to the markets of the East. At the start of the New Age, when France, England and Holland emerged as important partners on the Eastern markets, the deficit problems became even plainer, particularly since the Polish-Lithunian market showed large export surpluses in the Baltic area, and these had to be paid for with precious metals. Furthermore, when the sea route round the Cape became an alternative to the transit route via the Middle East to the Asian markets, there was an increased need for exports of precious metals from Western Europe. Thus during the 16th–18th centuries European world trade was faced with a constant balance problem on three markets: the Baltic area, the Levant and Asia. Very substantial sums were involved. Their magnitude is illustrated by the tables below, which set out the minimum figures for annual requirements in millions of rix-dollars (of 25.98 grammes of silver).

The old deficit area, the Levant, largely lost its significance as a transit region between Asia and Europe by virtue of the fact that from around 1600 onwards the Dutch and English East India companies effected what was essentially a reorganization of the trade routes between Asia and Europe. In fact a part of the transit traffic still continued to flow through the Levant, and the Levant's own trade with Europe showed large surpluses which were still settled by precious metals from Holland, France and England during the 17th and 18th centuries.

In the course of the European trade round the Cape to Asia, large quantities of precious metals flowed eastwards. On the Asian markets as a rule up to two-thirds of the purchases were paid for with precious metals. At times the companies were able to obtain precious metals in Asia - as the V.O.C. did in Japan in 1630–1680 - or in transit from Europe via the Middle East, but the bulk of the precious metals came round the Cape from Europe.

As far as the Baltic trade was concerned, there were large deficits which increased from the middle of the 16th century onwards. When the Dutch

Table 1
The bullion flow between Europe and the Levant
(In millions of rix-dollars per year)

Year	1600	1650	1700	1750	1780
Holland	0.6	0.8	1	1.5	1.5
France	...	1	1	1	1
England[1]	...	...	...	...	...
Total	1	2	2	2.5	2.5

1 Partly through Holland.

trade achieved growing importance and sharply curtailed the Hanseatic League's influence during the latter part of the 16th century, the expansion caused an increase in the demand for precious metals. Later, when the English trade became predominant in the Baltic area from the middle of the 18th century, there was a further increase in the demand for precious metals for the settlement of export surpluses in relation to Western Europe.

Table 2
The bullion flow round the Cape to Asia
(In millions of rix-dollars per year)

Year	1600	1650	1700	1750	1780
Holland	0.3	0.4[1]	2	3	3.5
France	-	-	0.5	2	2
England	0.1	0.4	1.5	2	2
Others (Portugal, Denmark, Sweden)	1	0.5	0.5	0.7	0.7
Total	1.4	1.3 [2]	4.5	7.7	8.2

1 Plus about 0.4 from Japan.

2 Plus about 0.4 from Japan.

However, the English deficit in the Baltic area was not paid for by direct exports of precious metals from England but was dealt with via Holland, which ultimately delivered precious metals to the Baltic market for payment of both Dutch and English deficits. In consequence the settlement of the balances in the Baltic area became mainly a Dutch concern.

Table 3
The bullion flow from Holland to the Baltic[1]
(In millions of rix-dollars per year)

Year	1600	1650	1700	1750	1780
Holland	2	2.5	2	2	4
Total	2	2.5	2	2	4

1 The figures given here for the Baltic area are largely minimum figures.

From the beginning of the New Age the increased demand for precious metals in world trade could be met out of supplies of silver and gold from the mines in America. In fact the transatlantic traffic supplied significant quantities of precious metals to Seville and Cadiz, and during the 18th century to Lisbon as well. Estimates of the sums involved appear in table 5, which also includes the figures from table 4 for the purpose of comparison.

Table 4
Estimated annual exports of precious metals from Europe to the East
(In millions of rix-dollars)

Year	1600	1650	1700	1750	1780
The Levant	1	2	2	2.5	2.5
The Baltic region	2	2.5	2	2	4
The route round the Cape	1.4	1.3	4.5	7.7	8.2
Total	4.4	5.8	8.5	12.2	14.7

A very large part of the imports of American precious metals into the Iberian peninsula were transferred by the European trade with Spain and Portugal to the main arteries of world trade in France, Italy, England and above all Holland. From the start of the 17th century Holland became to a great extent a clearing centre for the transfer of Spanish precious metals - and later Portuguese gold as well - to the deficit areas in the Baltic, the Levant and Asia. Thus, the large consignments of precious metals from America across the Atlantic to the Iberian peninsula were dispersed to a great extent via Western Europe to the deficit areas in the Baltic, the Levant and Asia. Precious metals also came to Asia from America via the Pacific Ocean and Manila.

It was the precious metals from America which facilitated the expansion of Western Europe's trade with the Baltic area, the Levant and Asia.
But the flow of American precious metals to Spain and Portugal was significantly greater in certain periods than the demand for bullion flowing from Europe to the East. Table 5 shows that such periods occurred around 1600 and 1750–1800. These points in time coincide with large price increases in Europe.

Table 5
Influx of precious metals from America into Spain and Portugal
(In millions of rix-dollars per year)

Year	1550	1600	1650	1700	1750	1780	1800
Influx into the Iberian peninsula	3	10	8-9	10.5-12.5	18-25	18-23	22-27
Bullion flow from Europe to the East	2-3	4.4	5.8	8.5	12.2	14.7	18

REFERENCES:

A. Attman. *The Bullion Flow between Europe and the East (1000–1750).* Gothenburg, 1981.

- - -. *Dutch Enterprise in the World Bullion Trade (1550–1800).* Gothenburg, 1983.

- - -. *American Bullion in the European World Trade (1600–1800).* Gothenburg, 1986.

AMERICAN TREASURE AND THE FLORENTINE COINAGE IN THE SIXTEENTH CENTURY

by Carlo M. CIPOLLA

We know where the American treasure came from and when it reached Spain. To find out where and when it went from there and the uses it was put to is an intriguing and complex historical problem. Part of the gold and silver was minted in the Colonies and part in Spain. How large a part, however, is not known precisely. The efforts by Ulloa give a partial answer to the problem but there is still much that remains to be known. Especially after 1555, part of the treasure left Spain. Again, how much and when we do not know. However, we do know that it left Spain in the form of ingots and coins, especially reales de a ocho, or pieces of eight. An Italian treatise of 1600 reports that the pieces of eight were more eagerly sought after than ingots because the former when crossing the various borders did not pay custom duties while the latter did. The same Italian treatise informs us that the pieces of eight were exported from Spain in boxes of 20,000 reales. Between 1550 and 1620 Genoa was perhaps the main trading and exchange center of Spanish silver in the south of Europe. When exported to Genoa by sea the cost of the operation as a percentage of the value of silver was as follows:

license of export from Spain	2.5%
insurance	2.0%
freight	0.1%
commissions	0.83%
miscellaneous	0.15%

Whether exported from Spain in the form of ingots or reales, when reaching its destination, American bullion was generally melted and transformed either into jewelry, silverware or local coinage. In fact the very same metal could undergo numerous successive transformations, being minted and reminted over and over again at the various European mints from one type of currency into another and into yet another again. For instance, during the 1560s and 1570s a large amount of the silver was brought to the Florentine mint in the form of reales. In the late seventies the mint noticed that the alloy of the new types of reales had deteriorated and consequently reduced the price it would pay for them. This measure reduced the inflow of reales while it increased the inflow of coins from Genoa, Naples and Milan which for the most part were made of American silver. Thus in the latter instance the same silver was successively transformed from reales or ingots into Genoese or Neapolitan or Milanese coins and then into Florentine coins. And this was not the end of the story because in its turn the Florentine currency could be exported and reminted somewhere else, let us say Venice.

From the above it follows that the figures relating to the outputs of the various mints cannot be added together without exaggerating the impact of the American metal. Still, the figures about the output of the mints can tell us a good deal about the flow of the American treasure and its dispersion throughout Europe.

Thanks to the research by F. Spooner and C.E. Challis we know a good deal about mint outputs in France and England during the sixteenth century. In Italy the main mints where those of Milan, Genoa, Venice, Florence, Rome and Naples. Besides these, there were a number of minor mints such as those of Ferrara, Mantua, Savoy, Bologna, Urbino, the independent feudal mints of the Genoese territory, etc. Unfortunately enough for most mints, including the major ones, the records relating to their output have been lost. For the mint of Genoa only a few scattered data have survived from which it is impossible to draw any conclusion. The loss is particularly lamentable because of the dominant position of Genoa as a financial market and the function of Genoa as a distributing center of the Spanish silver. As regarding Milan which was then a Spanish dominion, the original records of the mints' output have also been lost: but an eighteenth century numismatist saw the documents before they were destroyed and published series of data which, however, begin only with 1580. As for Venice, Tucci, who has done exhaustive research in the local archives has come to the conclusion that «no data are available for the first seven or eight decades of the sixteenth century». Florence represents the exception among the major markets of northern Italy.

For the Florentine mint scattered data are available for the first four decades of the century. They seem to indicate that while between 1503 and 1512 the value of the silver minted at the Florentine mint was higher than the value of the gold minted, between 1524 and 1532 the relationship was reversed and the value of the gold minted was above the value of the silver minted. Whether this change was related to the massive arrivals of gold from the America to Spain is matter of pure speculation. We are on safer grounds from 1543 because for the period 1543 through 1589 an abundant documentation is available on the activity of the Florentine mint and an almost complete series of data on the output of the mint can be reconstructed from the surviving documentation.

Table 1 and graphs 1 and 2 summarize my findings. I should point out that the graphs are constructed on different scales. Moreover, while the graphs

Table 1
Amount of pure gold and pure silver minted at the mint of Florence (yearly averages in kilograms)

Period	gold	silver
1543–1549	113	572
1550–1557	70	260
1558–1559	50	2.679
1560–1569	36	2.902
1570–1579	21	5.153
1580–1589	19	5.584

indicate output in terms of Florentine pounds of alloy, for the convenience of international comparisons the data of table 1 have been converted into kilograms of pure metal, that is, 24 carats for gold and 1000/1000 for silver. Both the graphs and the table indicate clearly that while the output of gold currency declined in the long run over the whole period of 1543–1589, the output of the silver currency increased suddenly and remarkably beginning in 1558 and in the two decades of the seventies and the eighties it reached respectively the average level of 5.2 and 5.6 tons of pure silver per year. To give a meaning to these figures suffice it to say that the Florentine output during those two decades represented respectively 5 and 3 percent of the silver that arrived in Seville. As to comparison with other Italian mints, it can be said that at the mint of Milan in the 1580s and 1590s the amount of silver coined reached the averages respectively of 12.7 and 9.0 tons per year.

To determine the level and the course of minting activity, however, raises more questions than it answers. Where did all those tons of silver come from? Directly from Spain or via Genoa and/or other places? When we find in the records that silver in the form of reales was brought to the mint by Spanish merchants we may assume that it came directly from Spain. When we find that it was brought to the mint by Genoese merchants we may assume that it came via Genoa. But such cases cover only a minor fraction of the silver inflow. Most of the silver was brought to the mint by Florentine merchants and bankers leaving us uncertain about the route followed by the metal. Moreover: why did the silver flow to Florence? In other words, in exchange for what? We cannot answer the question because we have no data about the Florentine balance of trade and of payments with Spain, Genoa, Naples and Milan. One thing should be noticed, however. Not only did issues

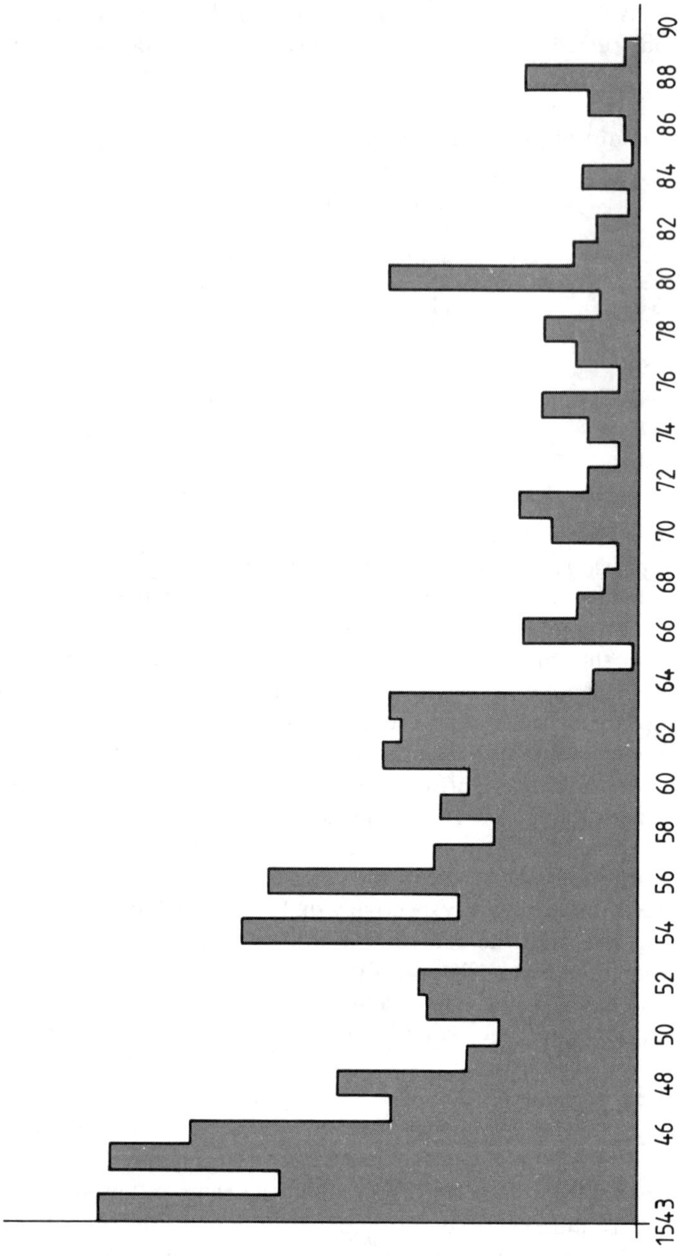

Graph 1. Volume of Florentine gold coinage expressed in Florentine ounces of gold at 22 carats.
One Florentine ounce = grams 28,295
Period covered : 1543 - 1589

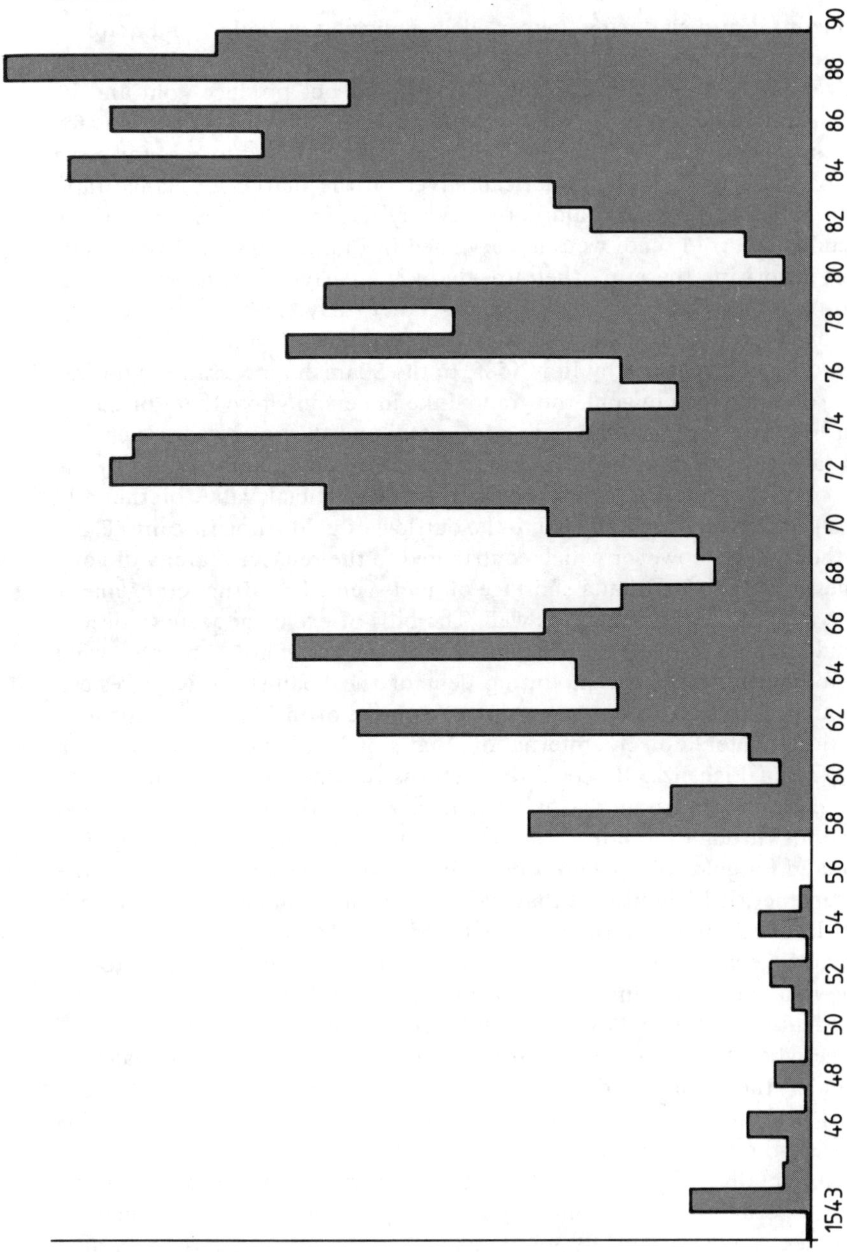

Graph 2. Volume of the Florentine silver coinage expressed in Florentine pounds of silver at 958,3 / 1000 alloy.

One Florentine pound = grams 339,5.

Period covered : 1543 - 1589

of gold coins diminish during the period, as indicated by table 1 and graph 1, all available evidence points to the fact that from the late fifties onward Tuscany was a net exporter of gold. Tuscany did not produce gold and its export meant depletion of its stock of gold currency. One of the reasons for this phenomenon was that the grand duke, a Medici, failed to recognize the effects of the arrival of the American silver on the market exchange ratio between gold and silver and did not allow any rise in the official rate of the gold scudo. The gold scudi were undervalued in Florence at the official rate. It is not surprising therefore that already by the early sixties it was increasingly difficult to find in Florence gold scudi to pay for bills of exchange drawn on Florence. By the seventies the shortage of gold currency had reached critical levels and by June 1585, to the Spanish ambassador who was trying to obtain a loan in gold, the grand duke in person wrote that «presently here in Florence one can find only a few scudi and they cost very much.»

As I said, the scarcity of gold in Florence was fundamentally caused by the refusal of the monetary authorities to raise the official value of the gold scudo, a failure which actually led to the outflow of gold from Tuscany. There were other factors however which contributed to the relative scarcity of gold, as witnessed by the fact that a shortage of gold seemed to afflict other major markets in Italy and in France as well. The bills of exchange were traditionally paid in gold. Now, from the middle of the sixteenth century, owing among other things to the mounting debt of the Spanish crown, Western Europe experienced a vast increase in the volume of international financial transactions centered on the international fairs of Medina del Campo, Lyon, Besançon and Piacenza. It seems that returns on financial assets tended to rise in relation to return on real assets. The increased volume of international transactions through the bills of exchange led to a proportionate increase in the demand for gold coins whose supply did not prove as elastic as that of the silver currency. It is significant that starting with the middle of the sixteenth century in one Italian market after another it was decided to issue big silver coins broadly equivalent in value to the gold scudo. Florentine sources testify that the issue of such coins in Florence in 1568 was due to the difficulty the bankers had in finding gold scudi on the market in order to settle the bills of exchange when they came to maturity. These new silver coins circulated in Italy under the name of ducatoni or scudi d'argento or piastre. Table 2 provides information about the numismatic characteristics of such pieces which were the Italian version of the pieces of eight and the thalers.

Fundamentally the second half of the century was for the Florentine currency a period of almost complete stability in terms of silver. The personality and the convictions of the grand duke played quite a role in maintaining this stability which he considered a matter of prestige and orderly governm-ent. The silver parity of the Florentine lira which was at 4.6 grams in 1550 was still at 4.5 grams at the end of the century. The stability in terms of silver,

Table 2
The earliest Italian silver Ducatoons

mint	year	weight (grams)	fineness (n/1000)	pure silver (grams)
Milan	1551	. . .	. . .	30.62 (?)
Venice	1563	32.90	948.0	31.19
Genoa	1567	37.27	958.3	35.71
Florence	1568	32.56	958.3	31.20

however, did not mean stability in terms of gold or in terms of commodities and services. As to gold the increasing discrepancy between the official value of the gold scudo and the market value of its metallic content, as we have seen, led not only to a reduction of the inflow of gold to the mint but actually to a net outflow of gold from the country. As to the prices of commodities and services, the index calculated by Parenti (an index which incidentally is overly weighted by agricultural products) shows a doubling of prices between 1550 and 1600. Within this broad upward trend there were long run ups and downs which show no relation to the movements of the output of silver currency. The phases were as follows:

1552 : 50
1560 : 71
1565 : 64
1575 : 86
1585 : 76
1590 : 78
1600 : 100

Here I wish to point to the phase of declining prices between 1575 and 1585 and the following phase of stagnating prices between 1585 and 1590: a good fifteen years during which in Tuscany the Price Revolution was brought to a halt. The reason for this was that during the period in question Florence experienced a severe banking crisis which had ominous consequences for the economy. The deflation that resulted despite the large output of silver currency warns us that our analysis would be both incomplete and incorrect if we were to limit it to the metals and metallic currency, ignoring the importance and the role of the bank deposits which were based on them.[1]

[1] For a detailed analysis of the facts sketched in this paper and references to the bibliography and the archival sources see my forthcoming book *La moneta a Firenze nel Cinquecento*, Il Mulino, Bologna.

PRECIOUS METALS, MONEY AND CAPITAL

by Michel MORINEAU

Due to a combination of circumstances, the main one being that the author cannot deliver this «paper» in person, the present contribution is no more than a summary of the subject and basically tentative in character. Even so, it should not be totally without merit to the extent that, drawing on the fruits of research already completed[1], it aims to shed light upon and give due weight to progress already made, to arouse the interest of all colleagues participating in the congress, primarily with a view to improving upon our text, and, finally, to discover what new avenues are opened or horizons glimpsed as a result.

1. Improvements to be considered

The task I set myself in the *Incroyables Gazettes ...* was to piece together the arrivals of treasure from the Americas in Europe (largely by way of Spain and Portugal), by drawing on a source which had not hitherto been exploited in a systematic fashion from our point of view, and which, upon use, turned out to be excellent, even if not quite perfect. The pattern of the arrivals of American treasure emerges somewhat altered and a certain number of views held in ignorance of the real facts either become completely irrelevant or require substantial revision. I shall not labour the point since anyone who wishes to follow the argument can refer to my book. I would only point out that, in addition to the theme in this chapter, my book also discussed the redispatch of money to the Orient, the building up of Europe's metal stock, the circulation of precious metals throughout the continent and various factors, such as the establishment of prices, which effected the mechanism of trade.

Back in 1980 I wrote that the curves presented in my work were the most accurate which could be posited at the time. This admission was not a bit of

[1] M. MORINEAU, *Incroyables Gazettes et Fabuleux Métaux. Les retours des trésors américains dans les gazettes hollandaises*, Paris-Cambridge, 1984–1985.

bragging - silly by definition - but based on an awareness of a fact and from an awareness too of the way knowledge advances in fits and starts and of the fact that it is sometimes necessary to reach beyond where we now stand. If in the four-year period between 1976 and 1980 it was possible to present a fairly substantial amount of new information in my study (which explains its structure and the appendices), it stands to reason that a future study should contain further improvements: this business of divining sources, of cross-checking, of criticism. I myself pointed out issues on which fresh thinking might be useful and I would like to recall them briefly:

1. the potential recovery of «doublets» in buildings;
2. the wrecks (?) of ships returning to the Mediterranean coast of Spain at the end of the eigtheenth century (especially at Barcelona);
3. a more meticulous examination of the trough between 1630–1650 concerning which it would seem information is to be gleaned, most probably, from documents other than the official register;
4. a reassessment of arrivals of Brasilian origin after 1760, now possible to trace from archival sources which became accessible after my book had been published;
5. an eventual readjustment of the general curve of accumulation in the light of a more accurate estimate of Europe's initial stock c. 1492.

Perhaps at this juncture it might be possible to outline a few epistemological considerations regarding the way in which history, and quantitative history above all, is made. Worth thinking about are: the trap presented by documentation which is too one-sided or which enjoys too celebrated a reputation (something which emerges from the gaps in the *Archiv de Indias* in Seville), and the lack of cooperation or constructive exchange of information, yet again. I think that historical research has much to gain from the «pooling» of knowledge and if I, for my part, am endeavouring to give extra conviction to my conclusions, I must confess that I hope to be able to do this in a joint effort with my colleagues.

2. Sure ground

An acknowledgement of potential improvements implies no admission of confusion on our part and even less does it suggest second thoughts about the work accomplished so far. It is obvious that one cannot claim in advance that certain expected results will not necessitate at least partial revision of conclusions already drawn. Were such a revision carried out (and justified), we would be the first to rejoice. Nonetheless, the margin of error we imposed upon ourselves and our constant concern that our facts be as firmly grounded as possible combine to make us optimistic and even quite positive on a fair number of propositions adopted.

By way of example, I should like to take the curve likely to give rise to the most lively debate: ie. The stock over three centuries, close to the right. Any reassessment of the initial stock (c. 1492), provided that it is on a fairly acceptable scale, is bound to involve an adjustment of the curve and, possibly, if the basic figure opted for is lower than that taken from Hume, the section of the curve corresponding to the 16th century would need to be redrawn. It would not *ipso facto* be necessary to return to the quantitative hypothesis (Bouillon) in order to explain the price use at the period. In fact the mechanism involved in this price use has been given independent consideration (see the introductory article) and this «relationship» *applies* in any case, turning up also in the 18th century. What might be worth having a fresh look at - or, perhaps, considering for the first time - would be the precise impact of the gold and silver from the Americas, Africa or the mines of Central Europe on the modalities of the price increase. Study of this issue was already accepted, suggested and indicated as desirable in the introductory article.

The same applies, it must be said, to the different effects caused by the demographic growth so dear to Hammarström, Wrigley and Schofield. It is impossible for me to enter into a discussion of this complex problem here but, if circumstances are more favourable in 1986 than they are now, it is a topic I hope to address in Bern[2]. Yet so as not to keep everyone in suspense, I would like to say the following. Once one has put the mirage of univocal hypotheses - whether they be quantitative or populationist - behind one, and returned to square one, it becomes evident that the essential elements of the problem have been just skimmed over. Take the market for example. What really happens on a market? Or on markets (for they are not all the same in character)? Where the different products are concerned? Halle de Chèrleville's study[3] recently taught me a great deal, but a lot remains to be done and the economists, for their part, have been scandalously slow in returning to first principles (at the moment, they are the best informed authorities). What appears to me beyond dispute at the present juncture is that the alluring development contained in Fisher's formula is an absolute essential and the key to numerous false dilemmas. The reorganization of the minting of money in eigtheenth-century France[4] completely altered once again the light in which the problem must be seen by highlighting an

[2] IXth International Congres of Economic History. 17th Section (J.J. McCusker): International Price History. The Next Generation.

[3] A. La Halle De Cherleville, *Fourniture des grains et mécanisme des prix,* in Actes du Congrès National des Sociétés Savantes, Reims, 1971 (reprinted in *Pour une histoire économique vraie*, Lille, 1985).

[4] J. Day (ed.), *Etudes d'histoire monétaire*, Lille, 1984.

unexpected «liberty» in the relationship between money and the general economy and a rather early dependance on politics. It is impossible to ignore this development.

Another notion which must be listed is that of capital. In the *Incroyables Gazettes* the traditional ratio between arrivals of precious metals and the formation of «capital» was seriously questioned. The argument put forward does not seem to me to have been understood, not even the facts on which it was based. Wallerstein responded to our assertions that Brazilian gold was neither «essential» nor «irreplaceable» for the rapid development of the British economy: «Morineau is tilting at windmills...». One is bound to express surprise at the (mischievous) way he is hard on himself. The real problem is, however, that one simple argument completely escaped him. A close look at English trade in the eigtheenth century shows that if it, quite logically, drew upon the sales possibilities offered by Portugal, and, through Portugal, Brazil in the days when it was riding high on gold, it did not suffer greatly from the subsequent change in trading patterns and had always, or at the same time, open to it outlets where there was no supply of precious metals (Ireland, the American colonies, etc. ...) which were just as important. Now, if it is true that gold and silver are purchased just like any other merchandise and are products which can be replaced by others, and are, therefore, «non-essentials», the same conclusion can be drawn as to their role as formers of capital. But, were we to develop this line of argument, it would take us too far because the very characteristics of precious metals and especially of gold - which, as has been repeated constantly since the days of Aristotle, is not the absolute standard of value - of capital, whose specific nature, effectiveness and metamorphoses have frequently been set aside, to the benefit of a categorical assertion, or rather of the incantation of a whirling dervish.

3. A desire for international cooperation

Wallerstein would have been able to draw a lesson different from the one we drew after studying the way in which the different countries acted vis à vis precious metals and the manner in which they conducted business in fonction of their stock.[5] He could have come up with circulation, the development of alternative means of payment, the choice of investments, the technological framework, sensitive or tactful approach to potential purchas-

[5] I. Wallerstein, *Système du monde du XVe siècle à nos jours*, Paris, 1983, p. 320. See also our *Incroyables gazettes*, studies 2 and 5.

ers, but all of these factors he has chosen to ignore. Simply compare England and France. Furthermore, it will be necessary to that other role for precious metals which has been brought to light, namely its role as financer of war. It is possible that by this shift we may have found a stimulant to economic activity. Yet the scheme for integration into the circuits of this activity cannot, for the moment, be identical to the scheme underlying the thought of most of the scholars currently writing about gold and silver.

By way of conclusion, I will again cite Wallerstein. Everyone has been well aware of the mix of data from sundry sources which he has used in his undertaking - an admirable undertaking for all that in terms of its aims and objectives. Admirable but, alas, not sticking to the rules of historical methodology we learned long ago. Wallerstein uses *all* authorities - at least all those whom he read (and since he is in any case omnivorous, that means everyone) but he fails to realize that all are not equally valuable nor are all valuable at the same time especially, as happens from time to time, if one contradicts the other. It is essential to observe a level-headed code of professional ethics and, as in all scientific disciplines, try also in history to cover virgin territory i.e. the unknown face of knowledge - including the area of statistics: in other words, it goes without saying that one has to proceed with caution, to recognize the margin of possible error, to move from knowledge of a certain weight to more convincing knowledge on the basis of additional information.

I have come full circle back to my original assertion. Yet, because I do believe that the initiative taken by the Stockton Colloquium is on the right track, I would like to transmit my very warmest wishes to the participants from my «exile» in Paris.[6]

---

[6] Translated from French by Dr. Patrick H. Daly.

PROBLEMS OF QUANTIFYING THE VOLUME OF MONEY IN EARLY MODERN TIMES. A PRELIMINARY SURVEY

by Joachim SCHÜTTENHELM

1. Complex of Problems

In the last few years historians focused their interest on monetary problems.[1] One of the major objectives is the size (and structure) of the money stock. The research in this special field is regarded as important mainly for three reasons:

- to write a more detailed, quantitative history of money for a certain country,
- to show and explain relationships between monetary growth and price level movements,
- to investigate changes in the money stock and their effect on economic growth and development.

For the last two reasons various models exist (mostly based on versions of the Quantity Theory of Money) in order to try to solve those problems.[2] But they are worthless for historians without the possibility of an empirical test. The obstacle is that an estimation of the quantity of money in prestatistical periods itself is generally an unsolved problem; only very few rough estimates exist.[3] The aim of this brief survey is to present three different methods of quantifying the volume of money. By contrasting the concepts, relationships are shown in the hope of stimulating more interdisciplinary research.

[1] E. VAN CAUWENBERGHE and F. IRSIGLER (eds.), *Münzprägung, Geldumlauf und Wechselkurse. Minting, Monetary Circulation and Exchange Rates. Akten des 8th International Economic History Congress (Section C 7)*, (Trierer Historische Forschungen 7), Trier, 1984.

[2] For the discussion of such models see D.O. FLYNN, *Use and Misuse of the Quantity Theory of Money in Early Modern Historiography*, in: E. VAN CAUWENBERGHE and F. IRSIGLER (eds.), *Münzprägung*, pp. 383ff.

[3] For example F.C. SPOONER, *The International Economy and Monetary Movements in France (1493–1725)*, Cambridge (Mass.), 1972; J.C. RILEY, *Monetary Growth and Price Stability: France (1650–1700)*, in: Journal of Interdisciplinary History, 15 (1984), pp. 235ff., used the data of Spooner for a research of economic relationships.

2. Definition of the Volume of Money

Before I present those concepts it is important to know what is included in the definition of money. With regard to long run development (Figure 1) the volume of money in early modern (and medieval) times can be defined as the stock of money in coined form. In these times metallic money was not the only form of money, but the major part of it. The following methods are restricted to this definition. But it must be stressed that they are empirically quite meaningless in later centuries, i.e. from about 1700 onwards, when fiduciary money started to forge ahead. The money stock -in the definition above- is expressed in nominal terms, i.e. in money of account. This requires that the (market) prices of gold and silver and also the rates of exchange be known.

3. Introduction of the Different Concepts

Here three concepts will be presented which are named after the chief source of data upon which they are based. We distinguish among the approaches of

1. precious metal-production,
2. mint-production,
3. coin finds.

Each method should ultimately lead to the same result: the volume of (coined) money. The concepts corresponding to the realistic stage of an open economy with foreign trade and loss of money are shown in table 1. Before I come to a more critical discussion (item 4) let me provide some brief comments on each concept.

3.1. Precious Metal-Production

With regard to the final destination the production of precious metals (gold and silver) can be divided into that part disposable for monetary means, i.e. for coins, and the remaining part for non-monetary purposes such as the manufacture of vessels or jewellery. But first we have to take into account losses which are irrecoverable; the most famous examples of such bullion-

FIG. 1 EPOCHS OF MONETARY HISTORY

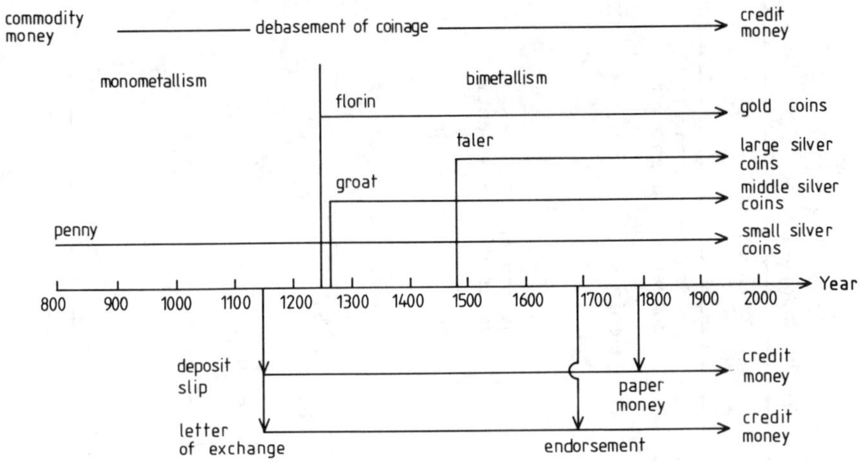

losses were those involving shipwreck. As to non-monetary consumption, there were also irrecoverable losses, e.g. the careless loss of jewellery. Despite this, the non-monetary portion remelted might play an important role by subsequently augmenting the coin stock. One of the most complicated problems involves quantification of bullion and coin imports and exports. As shown later on, these factors may considerably influence the volume of money. Besides, irrecoverable loss of coins in circulation occurs caused by wear and tear (constant loss) and sometimes by hoarding (variable loss).

3.2. Mint-Production

The concept of the mint-production seems to be clear; but we have to consider again the problematic factors of coin import/export and loss.[4]

---

[4] A dynamic version of this concept is given by E. VAN CAUWENBERGHE and D. HAENECAERT, *Mintage and Coin Circulation in the Southern Low Countries (14th-18th Centuries). Some theoretical reflections*, in: E. VAN CAUWENBERGHE and F. IRSIGLER (eds.), *Münzpragung*, pp. 169ff; see also the article of E. VAN CAUWENBERGHE and R. METZ in this volume.

Table 1
Concepts of Quantifying the Volume of Money in Early Modern Times

concept stage	1. precious metal -production		2. mint- -production	3. coin finds	
	+ production	+O		+ hidden and discovered domestic and foreign coins	$+M^{h}_{d,f}$
	- loss before subsequent treatment	$-L^{O}$			
open economy	+ import	$+Imp^{O}$		+ hidden and undiscovered domestic and foreign coins	$+M^{H-h}_{d,f}$
with foreign	- export	$-Exp^{O}$			
trade and loss	= production disposable for monetary and non-monetary means	O^{a}		+ domestic coins in circulation	$+M^{U}_{d}$
	- non-monetary consumption potentially disposable for monetary means	$-C^{a}$			
	- loss by non-monetary consumption	$-L^{C}$			

Table 1, continued

concept / stage	1. precious metal -production	2. mint- -production	3. coin finds
open economy with foreign trade and loss	= production disposable for monetary means O^M =	domestic mint-production M_d	
	+ import of foreign coins $+Imp^M$	+ import of foreign coins $+Imp^M$	+ foreign coins in circulation $+M^U_f$
	- export of domestic coins $-Exp^M$	- export of domestic coins $-Exp^M$	
	- irrecoverable loss of domestic and foreign coins $-L^M$	- irrecoverable loss of domestic and foreign coins $-L^M$	- irrecoverable loss of domestic and foreign coins L^M
volume of money (coin stock) =	M =	M =	M

The simple basic equations of the methods (in statistical terms):

1. precious metal-production
   $M = [(O - L^O + Imp^O - Exp^O) - C^a - L^C] + Imp^M - Exp^M - L^M$

2. mint-production
   $M = M_d + (Imp^M - Exp^M) - L^M$

3. coin finds
   $M = M^h_{d,f} + M^{H-h}_{d,f} + M^U_d + M^U_f - L^M$

3.3 Coin Finds

The concept of coin finds (single and cumulative coin finds, hoards) is from the methodical point of view the most complicated approach. Generally only a very small part of the coin stock, the hidden and then discovered coins are known. The basic problem is to connect this known part to the whole but unknown part, i.e. the domestic and foreign coins in circulation. This problem can be solved -under certain conditions- by the use of methods of random sampling. The method of calculation is presented below.[5]»

First the average percentage share π of a certain coin type k in a sample consisting of j = 1...n hoards is computed; the sample is constituted for the period t. This is:

$$(1) \qquad \pi_k = \frac{\sum_{j=1}^{n} G_{kj}}{\sum_{j=1}^{n} W_j}$$

with

π_k the percentage share of a coin type k in a sample constituted for the period t

$G_k$gzj the absolute value of the coin type k in a hoard j

W_j the absolute total value of all coins in a hoard j.

By using methods of probability theory π is related to the average percentage share of the considered coin type in the total stock (p) at the period t by the formula

$$(2) \qquad \pi = p$$

[5] For coin hoards and monetary circulation see generally Ph. Grierson, *The Interpretation of Coin Finds*, in: Numismatic Chronicle 7. Ser. 5, p. I-XVI (1965) and 6, p. I-XXI (1966), reprint in: Ph. Grierson, *Later Medieval Numismatics (11th-16th centuries)* , London, 1979. For methods of coin-find-analysis see H. Eichhorn, *Der Strukturwandel im Geldumlauf Frankens zwischen 1437 und 1610. Ein Beitrag zur Methodologie der Geldgeschichte*, Wiesbaden, 1973, pp. 186–354; P. Cerwenka and P.W. Roth, *Der Münzumlauf des 16.Jahrhunderts im Raume des östlichen Österreich. Ein Anwendungsbeispiel der elektronischen Datenverarbeitung in der historischen Forschung*, Graz, 1972, pp. 5–20; J. Schüttenhelm, *Der Geldumlauf im südwestdeutschen Raum vom Riedlinger Münzvertrag 1423 bis zur ersten Kipperzeit 1618. Eine statistische Münzfundanalyse unter Anwendung der elektronischen Datenverarbeitung*, Bd. I: *Grundlagen, Methoden und Darstellung*, Diss. Heidelberg, 1981 (is going to be published by the Kommission für geschichtliche Landeskunde von Baden-Württemberg in Reihe B Forschungen).

The total population, the volume of money, is then

$$(3) \qquad M = \frac{m_k}{\pi_k} \cdot 100$$

where m_k indicates the share of the selected coin type in terms of absolute value within the coin stock at the period t.

As can be seen the concept of coin finds needs no information about coin imports or exports; the hidden and then discovered hoards generally contain domestic and foreign coins together, which are used to estimate the real circulation.

4. Discussion of the Concepts

The difficulties which arise from each concept concern above all the availability of data and the methodical problems involved in the calculations.

Added to concept 1

- Production records are often incomplete or not available at all.
- There is still a lack of information about the distribution of precious metals; unfortunately there exist no reliable estimates of colonial bullion export in various European countries.[6]
- The concept itself gives no idea either about import and export of coins nor about coin losses.

Added to concept 2

- Mint-production data are also often incomplete or not available at all.
- Mint records provide no information about coin import or export. If import of metallic money exceeds export (vice versa), e.g. by a disequilibrium of the balance of trade, the money stock derived from mint output data will be underestimated (overestimated). The divergencies could be very large especially for countries with poor or no gold or silver mines. A good example of this situation is offered by the area of Southwestern Germany which is almost identical with the district of the «Schwäbische Kreis» (Figure 2). Calculations based on coin finds show that the percentage of the domestic specie in circula-

[6] For a review of production records and distribution of precious metals see: H. KELLENBENZ (ed.), *Precious Metals in the Age of Expansion. Papers of the XIVth International Congress of the Historical Sciences*, San Francisco, 1975 [= Beiträge zur Wirtschaftsgeschichte 2, Stuttgart 1981].

tion is small (Figure 3 and 4). In other words foreign coins, i.e. coins imported from other german districts and foreign countries, dominate.

- Furthermore the problem of (constant and variable) loss remains unsolved by the concept itself. Loss rates must be estimated by indirect methods.[7]

Added to Concept 3

- For many regions and periods there exist few wellinventoried coin finds.
- Calculations based on coin finds use methods of random sampling. But the reliability of a hoard (or a sum of hoards) as a representative sample of the real currency is never sure.
- Methods of coin find-analysis first lead to estimations of the structure of the currency, but not directly to the size of the money stock. Additional information about the mint output of at least one coin type must be known.
- Despite these difficulties only the methods of coin find-analysis indicate the share of domestic and foreign coins and also structural changes in circulation. Furthermore these methods permit computations of a global loss rate (covering the problems of constant and variable loss), even for different coin types.

Empirical results are presented for the area of Southwestern Germany. Table 2 shows the coin groups mentioned in Figure 1 and the special coin types which belong to each group. These were the most important coins in circulation in the 15th and 16th century. The structural changes in monetary circulation are illustrated in Figures 5 and 6. In the sphere of small and middle size silver coins the outstanding specie was the *Batzen*. This coin type, mainly struck by mints in Southern Germany, replaced the old groats (*Plapparte* and *Groschen*).[8] Moreover structural changes appear in the field of large silver and gold coins where the *Taler* takes the place of the *Gulden*. The percentage share of the two dominating coin types of the 16th century are shown in Figures 7 and 8. The curves are computed by application of methods of polynomial regression. The coinage act of 1535 brought an end to the minting of the *Batzen*. It is remarkable that simultaneously with the decline of this specie the rise of the *Taler* continued. In fact, after 1537

---

[7] See E. VAN CAUWENBERGHE and D. HAENECAERT, *Mintage and Coin Circulation*, pp. 152ff. and the literature cited there.

[8] For the Batzen see detailed: J. SCHÜTTENHELM, *Zur Münzprägung und Silberversorgung süddeutscher Münzstätten im frühen 16. Jahrhundert*, in: Montanwirtschaft Mitteleuropas vom 12. bis 17. Jahrhundert. Stand, Wege und Aufgaben der Forschung. Der Anschnitt, Zeitschrift für Kunst und Kultur im Bergbau, Beiheft 2, Bochum, 1984, pp. 159 ff.

FIG. 2 Districts ("Kreise") of the German Empire 1512 - 1806

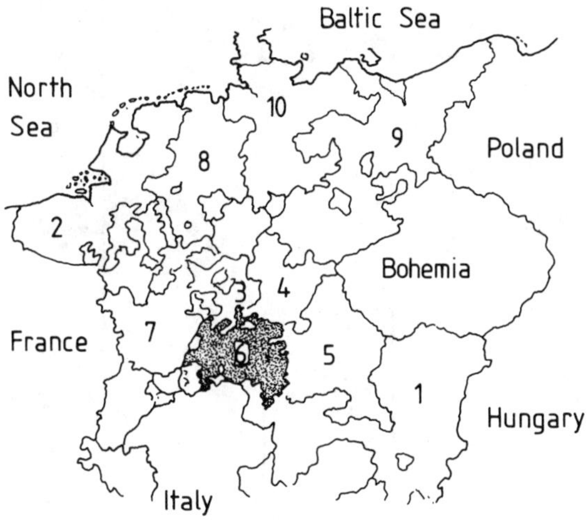

1	Österreichischer Kreis
2	Burgundischer Kreis
3	Kurrheinischer Kreis
4	Fränkischer Kreis
5	Bayerischer Kreis
6	Schwäbischer Kreis
7	Oberrheinischer Kreis
8	Niederrhein - westfälischer Kreis
9	Obersächsischer Kreis
10	Niedersächsischer Kreis

FIG. 3 Percentage share of domestic and foreign coin types (small and middle silver coins) in circulation in Southwestern Germany 1420 – 1620

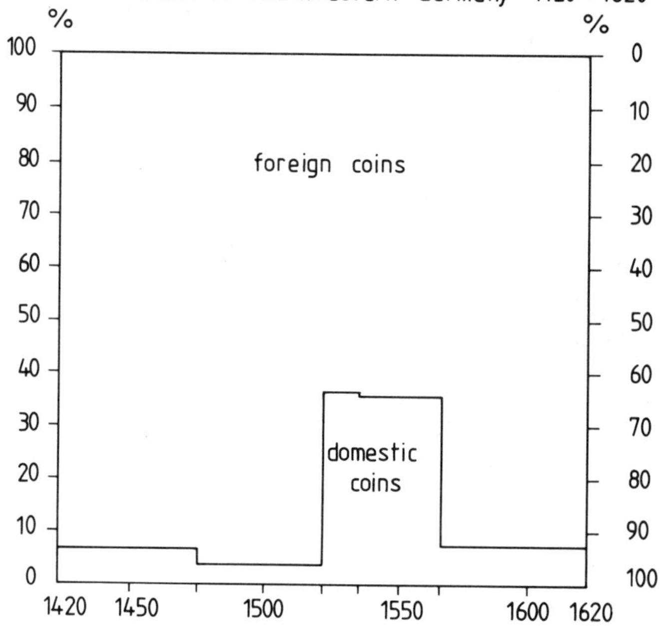

FIG. 4 Percentage share of domestic and foreign coin types (large silver and gold coins) in circulation in Southwestern Germany 1420 – 1620

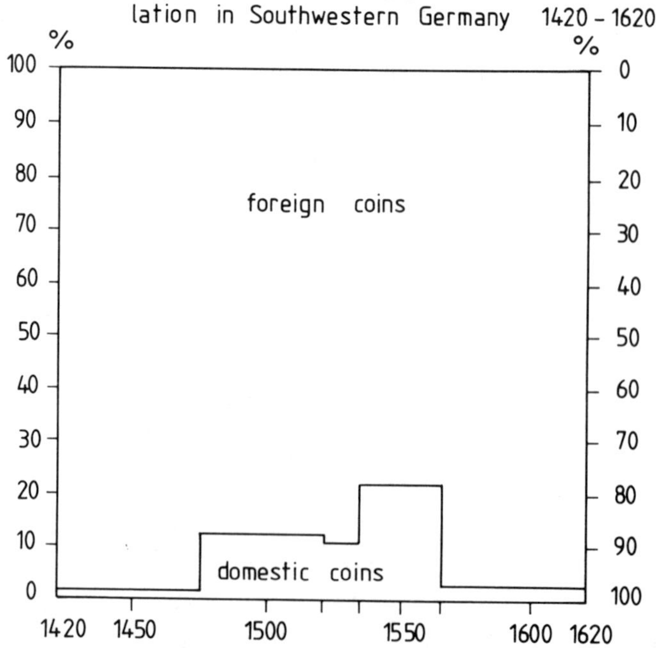

Table 2
Groups and Types of Coined Money

group			type
(1)	(2)	(3)	(4)
silver coins	small silver coins	1	Heller Pfennig Zweier
	middle silver coins	2	Dreier Kreuzer Halbgroschen
		3	Schilling Halbbatzen Zehner
		4	Plappart Groschen Batzen
		5	Sechskreuzer Halblire Zehnkreuzer Zwölfkreuzer
	large silver coins	1	Fünfteltaler Vierteltaler Teston Halbtaler
		2	Taler
gold coins	gold coins	3	Gulden
		4	Krone Dukat Doppeldukat Nobel

FIG. 5 Percentage share of groups of coined money (small and middle silver coins) in circulation in Southwestern Germany 1420 – 1620

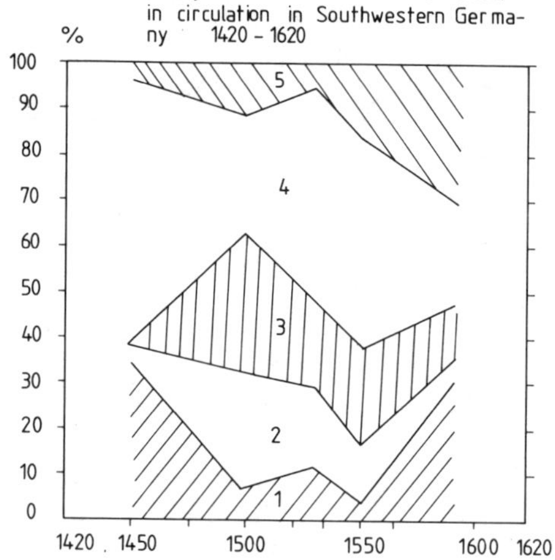

FIG. 6 Percentage share of groups of coined money (large silver and gold coins) in circulation in Southwestern Germany 1420 – 1620

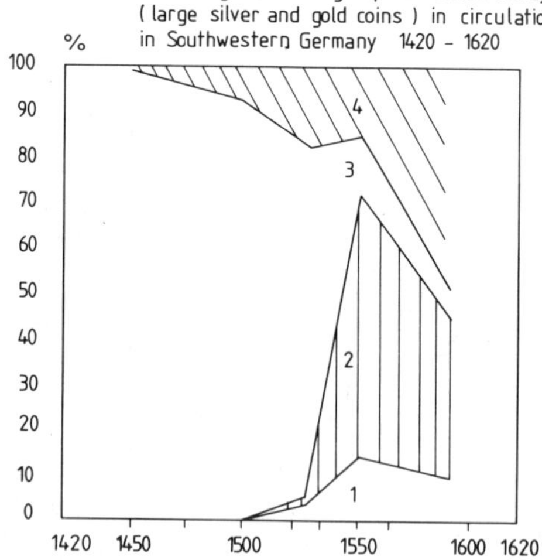

almost all South German mints started to produce *Taler*coins. More than 50 per cent of these large silver coins in circulation came from South German mints. It seems to be quite clear that the silver of the *Batzen* was used for the minting of the new *Taler* and other silver coins (the new *Halbbatzen* and *Groschen*) which came up after the middle of the 16th century. The loss rate of the *Batzen* (7,6%), shown in Table 3, gives an idea of the immense process of recoinage.[9] It is noteworthy to say that the average loss rate (3,2%) of all

[9] See the full computations in: J. SCHÜTTENHELM, *Geldumlauf*. (see n. 5)

middle silver coins - the most important group of coined money - correspond almost perfectly with the loss rate calculated by Van Cauwenberghe and Haenecaert (3,4%);[10] and these authors applied other methods than the methods of coin-find-analysis.

5. Conclusion

Which concept is the best? There can be no general answer to this question. The choice of one of these methods first of all depends upon the available data set of the country and time in question. Besides this pragmatic decision it must be stressed that none of the presented concepts allows in its pure form a satisfactory estimation of the volume of money. The most useful method seems to be the concept of mint-production, but it may lead to erroneous results if coin imports or exports reach relatively high magnitudes or if a standard loss rate is used in the computations. The weak points can be moderated by combining the study of mintoutput with advanced methods of coin find-analysis. A combined application of these two concepts saves bullion production records from worthlessness; if available, production figures may furnish a powerful instrument of control.

[10] E. Cauwenberghe and D. Haenecaert, *Mintage and Coin Circulation*, pp. 168ff.

FIG. 7 Percentage share of the Batzen in circulation of small and middle silver coins in Southwestern Germany 1420-1620 (based on coin finds)

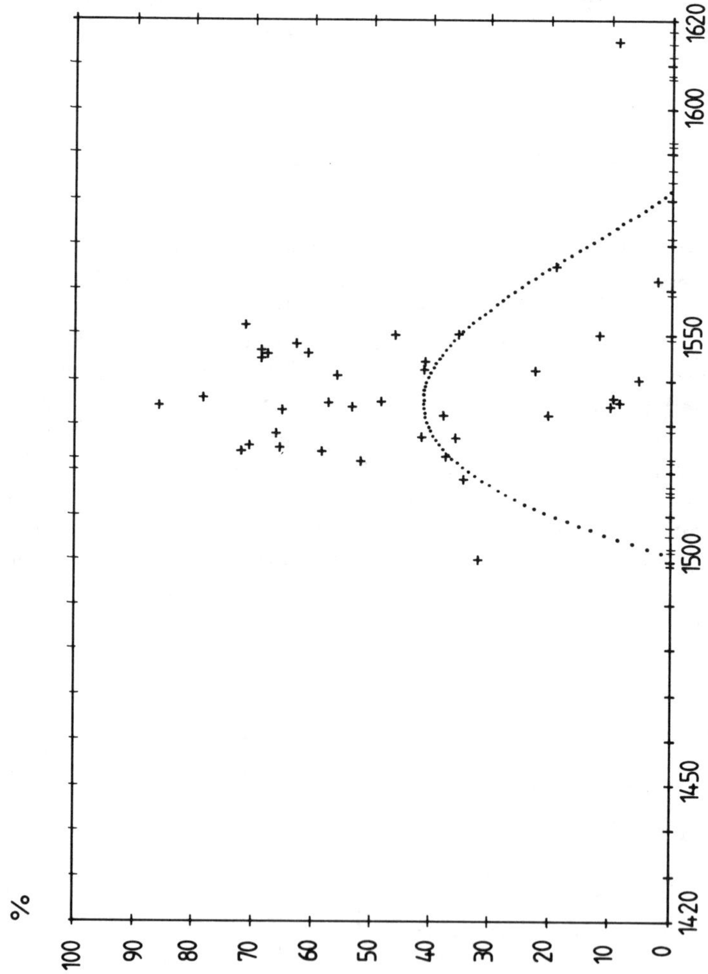

FIG. 8 Percentage share of the Taler in circulation of large silver and gold coins in Southwestern Germany 1420–1620 (based on coin finds)

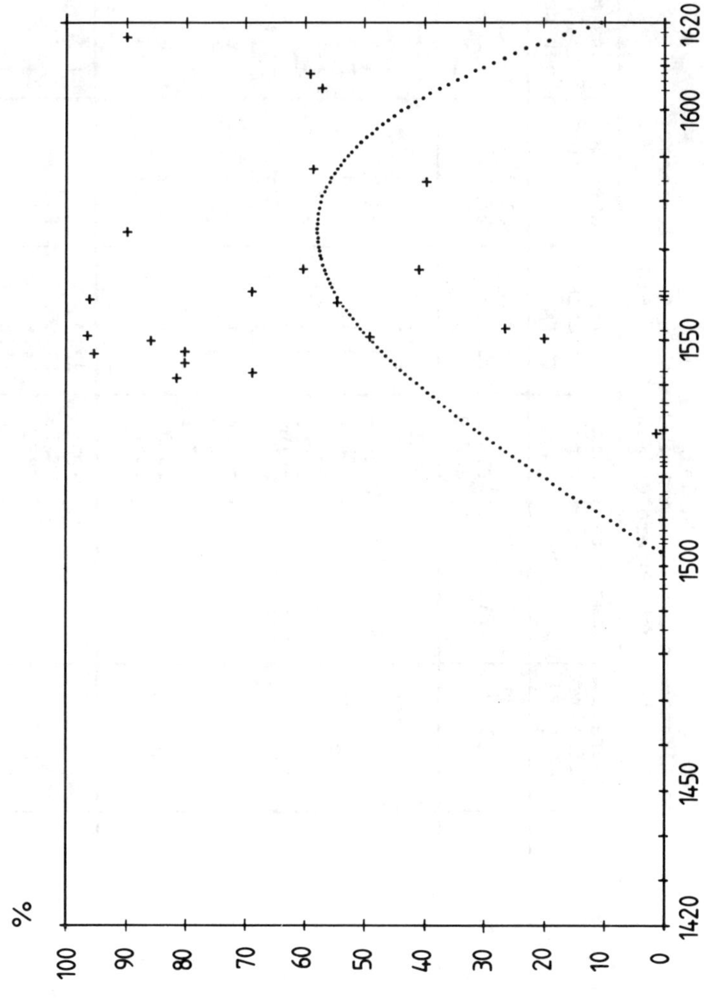

Table 3
The Lifetime and Loss Rate of Coins in Circulation in Southwestern Germany 1423-1618

group	half lifetime years	average lifetime years	loss rate per cent	coin type	half lifetime years	average lifetime years	loss rate per cent
small silver coins	23.5	27.1	4.3	HELLR	36.8	41.3	2.7
				PFENN	23.3	26.7	4.6
				ZWEIR	53.0	57.8	1.1
middle silver coins	22.3	27.8	3.2	DREIR	19.5	20.3	6.3
				KREUZ	35.5	36.4	3.8
				HGRSH	34.0	32.7	4.7
				SCHLL	60.1	55.4	3.0
				HBATZ	22.6	27.6	3.8
				ZEHNR	40.7	41.1	4.8
				PLAPP	61.0	64.2	1.0
				GRSCH	29.2	36.6	2.3
				BATZN	16.1	16.6	7.6
				6KRUZ	13.2	18.3	3.5
				HLIRE	53.8	50.4	2.1
				10KRZ	20.5	24.2	4.0
				12KRZ	17.3	23.1	3.8
large silver coins	9.0	11.7	6.0	FTALR	23.4	25.4	4.2
				VTALR	14.0	24.4	2.8
				TESTN	14.6	17.0	5.0
				HTALR	11.6	16.4	3.4
				TALER	8.8	10.8	6.1
gold coins	36.8	41.3	2.7	GULDN	25.4	30.0	3.9
				KRONE	23.3	33.2	2.3
				DUKAT	20.4	26.0	3.5

EXPORTS OF JAPAN'S SILVER TO CHINA VIA KOREA AND CHANGES IN THE TOKUGAWA MONETARY SYSTEM DURING THE 17th AND 18th CENTURIES

by Kazui TASHIRO

Japan, which produced huge amounts of precious metals from the late 16th century, exported silver to China in order to purchase Chinese raw silk and silk fabrics.[1] In the beginning, the Portuguese had a monopoly of that trade through Macau. After the isolation policy of the 1630s enforced by the Tokugawa government, however, the Dutch, who had succeeded in excluding their Western «trade rivals», actively exported silver to China via the Nagasaki-Taiwan route. Despite this, Chinese ships were carrying many times more silver from Nagasaki than the Dutch. However, from the 1670s on, exports of Japanese silver from Nagasaki declined drastically because the *shogunate*, amazed at the huge flow, prohibited its exportation by the Dutch in 1668, and exported copper instead of silver. According to the popular view, exports of Japanese silver were negligible by the end of the 17th century.[2]

In fact, however, an alternative route for the export of Japanese silver remained open for 60 more years, and silver kept flowing to China through it. This route ran from Tsushima, the north-westernmost part of Japan, to China through Korea.[3] At that time, most silver carried to China was discharged into China's southern coastal provinces, such as Fukien or Kwangtung. Instead, the Tsushima-Korea route ran north, through several cities in Japan, Korea and China, and ended at Peking.

---

[1] W. ATWELL, *International Bullion Flows and the Chinese Economy (Circa 1530–1650)*, in: Past and Present 95 (1982); K. YAMAMURA and T. KAMIKI, *Silver Mines and Sung Coins. A Monetary History of Medieval and Modern Japan in International Perspective*, in: J.F. RICHARDS (ed.), Precious Metals in the Later Medieval and Early Modern Worlds, 1983.

[2] S. IWAO, *Kinsei Nisshi Boeki ni kansuru Sūryōteki Kōsatsu* [Sino-Japanese Trade in the Edo period from a Quantitative Perspective], in: Shigaku Zasshi 62–11 (1953), p. 28; YAMAWAKI TEIJIRŌ, *Nagasaki no Tōjin Bōeki* [Sino-Japanese Trade in Nagasaki], Tokyo, 1964.

[3] Added to this, there was the route from Satsuma, the southernmost province of Japan, to China via the Ryukyu Kingdom. This trade is not discussed in this paper, but the Tsushima-Korean route, almost totally neglected by historians, was much more important quantitatively than the Satsuma-Ryūkyū route.

Apart from trade at Nagasaki, the Tsushima-Korea trade route continued to carry exports of silver to China from the mid-17th century on, and in return flowed imports of Chinese raw silk, silk fabrics and Korean ginseng to Japan until the mid-18th century. Japanese traders were able to continue exports of silver to Korea in those days despite the shogunate controls because the trade system was completely different from that at Nagasaki. Since all silver exported from Tsushima was silver currency issued by the shogunate, however, the trade was seriously affected by the changing monetary system in Japan after the late 17th century.

In this paper, I will focus on the flow of silver in the Japanese-Korean trade, and will show quantitatively how this flow affected commerce between Korea and China and the monetary system in Japan.

1. Japanese-Korean Trade in the Tokugawa Period

In the pre-modern era, the island of Tsushima, in the Korea Strait (see Figure 1), was like a «bridge» connecting Japan with the Eurasian Continent. Taking advantage of its geographical location, the islanders made their living through trade with Korea. The So family, territorial lords of Tsushima, obtained political control over the island during the 14th century, and by the early 15th century they monopolized diplomatic and economic relations between Japan and Korea. This arrangement was interrupted by Japan's abortive invasion of Korea in the 1590s, but it resumed in the 17th century when the So restored relations with Korea for the Tokugawa shogun in 1607, and itself exchanged a diplomatic agreement with Korea in 1609. Thus, the So family's position as an intermediary was recognized by both the Tokugawa shogu nate and the Yi dynasty of Korea.[4]

The beginning of the Tokugawa isolation policy in the 1630s struck a serious blow at many Japanese engaged in overseas trade. The role of the So family, however, increased in importance. Basically, the Tokugawa policy of isolation had two roles to play with regard to the external world: the first, vis-a-vis the West, to expel Catholicism, and the second, vis-a-vis East Asia, to make clear Japan's independence from the Chinese world order - the Sinocentric community of pre-modern East Asia. As far as the latter was concerned it permitted only private trade with the Chinese in Nagasaki, and rejected

[4] T. Tanaka, *Sakoku Seiritsuki Nitchō Kankei no Seikaku* [The Nature of Japan-Korea Relations in the Early Edo Period], in: Chōsen Gakuhō 34 (1965); H. Nakamura, *Nissen Kankeishi no Kenkyū* [A Study of the History of Japan-Korea Relations], 3 vols., Tokyo, 1969.

any diplomatic relations with the Chinese Government. Japan thus avoided joining the Chinese world order, while maintaining diplomatic relations with Korea as an equal partner and with the Ryūkyū Kingdom as a Japanese tributary.[5] The So family kept its unique position in East Asia throughout the Tokugawa era, and the shogunate regarded the Tsushima-Korea trade as important because it was one of the few trade routes for supplying Chinese goods to the Japanese domestic market.[6]

In contrast to the Japanese traders in Nagasaki who did not leave Japan, Tsushima islanders did travel for trade at the Japanese concession in Pusan, the southern tip of the Korean peninsula, in boats under the So's control. These were voyages specifically for overseas trading only during the Tokugawa era. Shipping from Tsushima was controlled by passports which the So issued, and was limited to 20 boats a year by the diplomatic agreement of 1609. However, over the years the Tsushima domain, with the acquiescence of the Yi government, managed to increase the number of boats beyond the quota so as to add to its profits. The domain developed a number of techniques to avoid the quota. For example, a number of small boats accompanied the officially approved mother ships on their two yearly trips; other ships and boats, on the pretext of carrying a «special mission» or a «letter to the concession,» made several trips a year. According to So documents, in 1709 90 boats came and went between Tsushima and Korea among which 30 or so of the smaller ships were carrying silver to the Japanese concession at Pusan.[7]

Tsushima-Korea trade took two forms: official trade, in which the Tsushima domain and the Korean government exchanged gifts and commodities of a kind and in amounts stipulated by the diplomatic provisions; and private trade, in which the Tsushima domain and merchants traded with the Korean officers and merchants at markets held 6 times monthly. Unlike the former, the latter was not subject to any restrictions, especially in exports of silver and imports of Chinese goods. This trade brought huge profits for the Tsushima domain.[8]

Although there is very little material concerning the private trade with Korea, an account book of the trade, recorded by Tsushima officials, does exist among the So family documents[9] and gives a record of the period from

[5] K. TASHIRO, *Foreign Relations During the Edo Period: Sakoku Reexamined*, in: Journal of Japanese Studies 8-2 (1982); R. TOBY, *State and Diplomacy in Early Modern Japan*, Princeton, 1984.

[6] K. TASHIRO, *Kinsei Nitchō Tsūkō Bōekishi no Kenkyū* [Studies on the History of the Diplomatic Relations and Trade between Japan and Korea], Tokyo, 1982, pp. 433–447.

[7] Ibid., pp. 202–218.

[8] K. TASHIRO, *Tsushima Han's Korean Trade (1684–1710)*, in: Acta Asiatica 30 (1976), pp. 85–86.

[9] There are 5 major collections of the So family documents as follows; National Diet Library (Tokyo), Historiographical Institute University of Tokyo (Tokyo), Keio Uni-

1684 to 1710. According to this, as is shown in Figure 2, silver comprised half of the annual total exports to Korea. In exchange for this silver, Tsushima islanders imported much Chinese white raw silk. As is shown in Table I, at the peak, in 1686, over 10 t. of silver currency was exported. The total amount exported from 1684 to 1710 amounted to about 188 t., an average of about 7 t. a year. During this same period, due to shogunate restrictions, the major commodity for export in Nagasaki was not silver but bar copper. Thus in 1686, for example, Chinese ships which were still permitted silver exports carried only 2 t. of silver from Nagasaki, one-fifth of the Tsushima-Korea trade. In 1708, moreover, the shogunate decided to fix an annual limit to silver exports of 0.6 t. in Nagasaki.[10] In other words, shogunate restrictions led to a economic reversal in the relative importance of the Nagasaki and Tsushima routes to the Japan-China silver trade.

2. The Changing Monetary System and Silver Exports

Tsushima's exports of silver to Korea created serious problems for the shogunate, both quantitatively and qualitatively, because all the silver was currency issued by the shogunate. The most important problem was caused by the recoinage in 1695 on, which had a great impact on the monetary system of the time.

Basically, Tokugawa Japan had a bimetallic standard system of gold and silver. In 1601, the central government, later to become the Tokugawa shogunate, issued gold coins (84% fine Keichō gold). These were circulated by *tael* (piece-by-piece) in East Japan. At the same time, silver currency (80% fine Keichō silver) was minted by the government and used in terms of «money by weight» in West Japan.[11] At first, the silver for trade was exported in the form of pure silver ingots. After the establishment of the monetary system, however, traders began to cast an eye on the high quality of silver currency. Adding to this, since the shogunate prohibited exports of pure silver ingots in 1609, a huge amount of silver currency flowed out of Japan.

About 100 years later, the then Finance Minister of the shogunate, Ogiwara Shigehide, reduced the quality of currency and increased the number in circulation. There were several reasons why he did this: a lack of

versity Library (Tokyo), Rekishi Minzoku Shiryōkan (Tsushima) and National History Compulation Committee of Korea (Seoul).

[10] YAMAWAKI, *op. cit.*, pp. 213-214.

[11] E. HONJŌ, *Economic Theory and History of Japan in the Tokugawa Period*, Tokyo, 1943, pp. 50-54; H. TAYA, *Kinsei Ginza no Kenkyū* [A Study of the Silver Mint in the Tokugawa Period], Tokyo, 1963, pp. 145-196.

currency resulting from the amount exported, an increase in the demand for the money which accompanied economic development, and a decline in mining productivity. Eventually, as Table 2 shows, his recoinages led to a drastic reduction in the quality of silver currency, which produced high inflation.[12]

The succession of debasements meant continual changes in the type of coin which the Tsushima domain was exporting. In 1697, Genroku silver (64% fine) in place of Keichō silver was minted by the shogunate, but the domain kept exporting Keichō silver for two more years for fear of introducing confusion into the private market. Meanwhile, Tsushima officials negotiated with Korea for the acceptance of Genroku silver since Keichō silver was in poor supply - as the saying goes, «bad money drives out good money». It took two years to make the Koreans accept Genroku silver because their poor melting techniques for precious metals made it difficult to fix an exchange rate between the old and the new.[13]

As already noted, Genroku silver had a purity rate of 64%. The exchange rate with Keichō silver should therefore have been at a premium of 25% (80/64). In the lack of any official announcement about the purity of Genroku silver, the Tsushima domain, which had been running some silver mines on the island for centuries, determined the exact purity by itself and imparted this information to Korea. However, because the Korean officials could not produce perfectly pure silver with their melting techniques, their first evaluation was that the silver content of Genroku silver was only between 60% fine and 62%, and thus should be at a 33% (80/60) to 29% (80/62) premium. After several negotiations and purity testings, it was finally decided that Genroku silver was «63% fine» and at a «premium of 27%» (80/63) compared with Keichō silver. Without doubt, this was a disadvantageous contract for the Tsushima domain. However, since long drawn out negotiations only stagnated the private market, the domain was obliged to make a contract with Korea as early as possible.[14] The Tsushima islanders began exporting Genroku silver to Korea, in 1699, but despite the high premium it had a bad reputation in the market. Moreover, as is shown in Table 2, after 1706, the Finance Minister again debased the silver currency, and finally, in 1711, issued Yotsuhō silver (20% fine), the worst silver currency in Tokugawa history. The Korean government and merchants were no longer interested in low value silver, such as the Yotsuhō silver. Inevitably, the Tsushima domain had to continue exporting Genroku silver, which

---

[12] TAYA, *op. cit.* pp. 168–169.

[13] SANG-UN CHUN, *Hanguk Kuahak Kishulea* [Science and Technical History in Korea], Seoul, 1983, pp. 270–273; K. TASHIRO, *Kinsei Tsushima-han ni okeru Nissen Bōeki no Ichi Kōsatsu* [A study for Japanese-Korean Trade by the Tsushima Domain], in: Nihon Rekishi 268 (1970), p. 94, p. 112.

[14] K. TASHIRO, *op. cit.*, pp. 94–95.

gradually became scarce as had Keichō silver. Thus, as Table 1 shows, total exports of silver in 1710 had dropped to about 2.3 t. of Genroku silver.

After negotiations that lasted for five years, in 1710 the Finance Minister finally permitted the minting of 5.3 t. per year of special quality silver (80% fine) and the exporting of it. Moreover, the Minister not only permitted the exchange of bad currency for the special quality silver with *no premium*, but also did not charge Tsushima for the expenses of minting it. The shogunate took these costs upon itself because it expected that the special silver would promote imports of ginseng, treasured as a medicine, from Korea. In fact, although Tokugawa Japan was experiencing a great boom in the demand for ginseng,[15] during the devaluation period it was in short supply due to the decline of silver exports. The special silver was even called the *ninjindai-ōkogin* (special silver to buy ginseng).

Although exports of the special silver did not begin until 1712 because of the time taken for diplomatic negotiations with Korea, its quality enabled the trade to survive the crisis caused by the successive debasements. Fortunately, in 1714 the shogunate restored the quality of money. Silver currency (Kyōho silver) remained 80% fine for 2 decades thereafter and trade gradually returned to the right lines. No account book of private trade survives for this era, but I can show clearly the quantity of silver carried to the Japanese concession in Pusan by the diary kept by head officials of the concession for over 184 years. The part of Figure 3 based on the period from 1711 to 1751 shows that the quantity of exports increased the year immediately following release of the special silver. Thereafter amounts again declined for a few years, only to rebound soon after. It also appears that the figures were far bigger than those for exports from Nagasaki, which totalled only 1.4. t. in the period 1714–1738.

During the period 1714–1738 when Kyōho silver was being issued, the amount of currency was reduced and there was a large reduction in the amount of silver currency in circulation. This caused a serious monetary shortage in West Japan, where silver currency was used and which was more developed economically, than East Japan, where gold coins were used. It is true that the exceptionally good quality of Kyōho silver helped cause the money shortage. In order to solve the problem, therefore, in 1730 the shogunate permitted the issue of paper money in several domains,[16] mainly in West Japan, and in 1738, the government again recoined to increase the amount of currency through debasement.

---

[15] T. IMANURA, *Ninjin shi* [The History of Ginseng], 7 vols., Seoul, 1934–1940.

[16] The issue of paper money in domains was prohibited by the shogunate in 1707. The repeal in 1730 was because of the shortage of silver cash. See, A. KOHATA, *Nihon no Kahei* [Money in Japan], Tokyo, 1958, p. 179; and TAYA, *op. cit.*, p. 286.

However, I would insist that the high grade of silver currency was not the only reason for the money shortage. There was another reason as well; an increase in the exporting of silver cash. Table 3 shows the quantity of Kyōho silver carried to Korea by Tsushima while it has being issued by the shogunate. As is evident from this Table, the Tsushima domain exported, at the minimum, 86.4 t. of Kyōho silver, which is equivalent to 6.9% of all the silver minted during that time. Moreover, if one takes into account the years for which records have not survived, the figure would probably reach 8%; and if the amount of silver exported along the Satsuma-Ryūkyū-China routs is added (which is not discussed in this paper) it would reach about 10%. Given these facts, as discussed above, there seems little question that the export of great quantities of silver had a huge impact on the circulation system in periods of money shortage.

Setting aside the question of why exports of silver rapidly declined after the mid-1730s, let us turn to the movement of silver from Japan to Korea, and from Korea to China.

3. The Seasonal Cycle on the «Silver Road»

The Tsushima domain handled its transport of silver by the boats very carefully. These would start from Kyoto - the main marketing center for Chinese raw silk in Japan - where the Tsushima domain sold the silk it had bought in Korea and collected the silver it would export in return. The journey from Kyoto to Tsushima lasted about 7 days. At Tsushima the silver was unloaded and shortly after, carried to the Japanese concession in Pusan by other boats suitably reserved for the carrying of silver.

These special boats were sent to the Japanese concession in groups of 2-3, or 5-6, and the silver was passed on to the Korean merchants in private trade. There was a delay of several months after delivery of the silver before Tsushima received any commodities in return. As already noted, the diaries which the heads of the Japanese concession at Pusan kept over a long period of time provides us with valuable information on the operation of the boats. I would like to use these diaries, and to present some interesting data about the amounts of silver transported each month, according to the solar-lunar calendar from 1711 to 1750. Table 4 clearly shows that the amount fluctuated each month, and that there was a certain seasonal rhythm. For example, in the period 1711-1720, the amount of silver transported in the 8th, 10th and 11th months was more than that in other months, and over the whole period covered by the Table, the total ratio of silver transported annualy went up to 15.8% in the 8th month, 21.3% in the 10th and 13.0% in the 11th. Taken together with the 7th month (9.5%), the amount would equal about 60% of the totals.

This seasonal cycle was a result not of natural conditions, but was a phenomenon specific to silver transport. This becomes clearer if one looks at the following instructions given by the Tsushima domain to their officers in Kyoto in 1681[17]: first, send 2.6 t. of silver to Tsushima in the 6th month for the proffering ceremony of the Chinese calendar, second, send 5.6 t. of silver to Tsushima in the 8th month for the Winter Solstice ceremony. The former had to be carried to Korea during the 7th and 8th months, the latter during the 10th and 11th months. In other words, the seasonal cycle of silver transporting began in the supplying stage. Why was the domain so concerned to carry silver during these specific seasons?

«The proffering ceremony of the Chinese calendar» and «the Winter Solstice ceremony» which appear above were events which deeply affected diplomatic and economic relationships between Korea and China. Every year the Yi dynasty dispatched missions to the Ch'ing court in fulfillment of its tributary obligations under the Chinese world order. After the mid-17th century, annual missions went regularly to Peking twice a year, to attend both the Winter Solstice and the New Year's celebrations, and to receive the Chinese imperial calendar. Adding irregular missions, it is estimated that the Korean government dispatched missions on an average of 3-4 times a year. Although the official membership of each mission basically comprised less than 35 people, the total number expanded to about 300 people during the Winter Solstice mission. This was the largest one, because the retinue, stable keepers, and odd-jobbers joined in the party. Most of them were merchants, engaged in Sino-Korea trade, not only officially but privately as well. Although it took the form of a diplomatic mission, considering the constituent members and their economic activities, it only appeared to be a «caravan». This «caravan» brought goods which the Tsushima domain wanted from China and returned carrying mostly Japanese silver.[18]

While Japan began to produce a huge amount of precious metals after the late 16th century, thanks to a mining boom, Korea produced very little; rather she chose to close down her mines and to exclude precious metals from the list of the tributary goods sent to the Chinese emperor. However, her location sandwiched between China and Japan ensured that she was never indifferent to silver: on the one side, she had the world's largest importer of silver, on the other side, the greatest producer of silver in Asia. In addition, because Korea had no valuable goods other than ginseng, it was as a result of transit trade that silver imported from Tsushima flowed to China.

[17] *Nagasaki Kenshi* [The History of Nagasaki], II, Tokyo, 1964, pp. 615–616.

[18] M. Ejima, *Tensō Nenkan ni okeru Chōsen no Saihei ni tuite* [Sino-Korea Annual Tributary System During 1627–1643], in: Shien 101 (1969), pp. 39–62.

Indeed, a considerable amount of silver was re-exported from Korea to China in several ways: for example, through private trade by both Korean messengers and merchants, and through smuggling on the Sino-Korea border area, mostly at the time of the entrance and exit of the Korean missions. At first, official members of embassies were permitted to carry ginseng for their private trade, thereafter they were allowed to carry about 2,000–3,000 *taels* (75–113 kg.) of silver or ginseng per person. Therefore, in the case of the Winter Solstice mission, which had 35 official members, it was possible for them to carry a total of 74,000 *taels* (about 2.8 t.) of silver in all. Also, when private trade in silver as well as ginseng became allowed, most of the official members entrusted their privileges of carrying out silver to the trading merchants, and a fixed amount of silver export was no longer observed. It is said that one Winter Solstice mission carried a total of 80,000 *taels* (3 t.) or more of silver.[19] Moreover, silver was also exported by irregular missions and by smugglers, though it is impossible to make an estimate of the amount.

The movements of the regular missions were fixed: the mission for the acceptance of the Chinese calendar started from Seoul in the 8th month and returned from Peking in the 12th. The missions for the Winter Solstice and New Year's ceremony (a much larger group) left in the 11th month and did not come back until the 4th month of the next year. These undoubtedly had an effect on the seasonal cycle of silver transport from Kyoto, because silver transactions in the Japanese concession became more and more active as the start of a mission approached. To summarize, Japan's silver flowed to China via Korea on the «silver road» - Kyoto-Tsushima-Pusan-Seoul-Peking - where there was a twice yearly surge in conjunction with the movements of the Korean missions. We can call the 2 main as the «Summer flow» and the «Fall and Winter flow» respectably.

1. Summer flow:
   Kyoto → Tsushima in the 6th month; Tsushima → Korea between the 7th and 8th months; Korea → China after the 8th month.
2. Fall and Winter flow:
   Kyoto → Tsushima in the 8th month; Tsushima → Korea between the 10th and 11th months; Korea → China after the 11th month.

[19] SEUNG-JU LIU, *Choson Hugi tae Chong Muyok ui Chongae Kuajong* [Development of Trading with Ch'ing in the Latter Yi Dynasty Period], in: Paeksan Hakpo 8 (1970), pp. 327–398; TS'UN-WU CHANG, *Ch'ing Han Tsung-fan mao-yi* [Sino-Korean Tributary Trade, 1637–1894], Taiwan, 1978.

4. National Economy and Silver Export

Lastly, I would like to discuss the reasons for the rapid decline in the high level of silver exports for Tsushima after the mid 1730s.

The trade between Tsushima-Korea had 2 particular characteristics. The first was that the economy of the Tsushima domain depended on the export of silver because it lacked an agricultural base. Since the amount of trade was determined by the volume of silver exports, a decline in the latter meant at immediate fall in domain revenue. Moreover the Tsushima-Korea trade financed the diplomatic activities between Korea and the Tokugawa government for which the domain acted as an intermediator. This meant that any fall in trade also carried with it the possibility of a breakdown in the diplomatic activities between the two countries.

The second concerned the management of the Tsushima-Korea trade itself. As noted above, despite Japan's restrictive trade policy overall, trade took place on the Korean peninsula. The shogunate could not actually watch the activities of the Tsushima domain at the market, and could not monitor the exact amount of silver exported. In other words, the amount exported depended upon market business conditions. Even if the shogunate tried to limit the trade, unlike in Nagasaki it was not able to tell whether the domain had obeyed the law. It was because of these two characteristics that the shogunate allowed such exports of silver to continue for so long.

However, a Confucian scholar, Hakuseki Arai, who acted as adviser to the 7th Shogun eventually reversed the former policy toward silver exports in favor of hitherto unconsidered «national interest». The adviser first made estimates of the production, stock and exports of precious metals in Tokugawa Japan, and in 1714 restored the quality of the currency. Taking a broad view beyond the framework of the Tokugawa political and social system, he argued that Japan should reduce exports of valuable treasure, especially silver and copper, and should instead promote domestic industries and raise productivity. Although the currency was devaluated again later, the shogunate basically supported his ideas.

As a consequence of this, the shogunate decided that at some point they had to bring an end to Tsushima's high exports of silver. Eventually, in 1737, the shogunate had an opportunity to do so. The very year before, the 8th Shogun, Yoshimune Tokugawa, had debased the gold and silver currencies, reducing the purity of silver currency from 80% to 46%. According to precedent, the shogunate permitted Tsushima to mine and export 5.3 t. of special quality silver. But at the same time, it laid down the condition that Tsushima had to pay a premium of 99% in order to exchange currency for the special silver, except in the case of silver to be traded for ginseng imports which was at a premium of 50%. It is unlikely that in practice the special silver was ever used for the importing of ginseng because the Tsushima domain could never afford a premium of 99%. The shogunate thus took the power necessary to control the silver exports of the domain at the supply end, and was able to monitor the actual amount of ginseng imported because the amount of special silver permitted at a 50% premium was based on the amount imported in the preceding year. The shogunate moves had a definite effect and exports of silver currency from Tsushima began to fall as the result of a decline in Korea-China trade.[20] This is the reason for the rapid decline after the mid-1730s, as is shown in Figure 3.

Meanwhile, in order to continue diplomatic relations with Korea, the shogunate had to protect Tsushima from the consequences of its economic failure. It therefore gave financial help, in the form of grants or interest free loans. It is estimated this financial help reached a total of 1.5 million *ryō* of gold cash (corresponded to 8 million bushels of rice) from 1734 to 1841. This money was presumably in compensation for the closing down of the «silver road».[21]

---

[20] K. Tashiro, *Kinsei Nitchō*, pp. 338–339.
[21] *Ibid.*, pp. 394–395.

Table 1
Silver Export to Korea (1684-1710)

year	silver cash[a] (kg.)	pure silver (kg.)	per cent of total exports
1684	7,267	5,813	66%
85	7,527	6,021	64
86	10,827	8,662	69
87	7,665	6,132	59
88	9,327	7,461	63
89	7,480	5,984	50
90	8,366	6,693	47
91	10,239	8,191	47
92	9,139	7,311	52
93	8,528	6,822	41
94	9,671	7,737	46
95	9,185	7,348	47
96	9,149	7,319	49
97	9,018	6,230	40
98	5,250	3,360	53
99	7,425	4,752	68
1700	5,868	3,756	98
01	10,237	6,552	64
02	6,776	4,336	64
03	2,737	1,752	59
04	5,062	3,240	70
05	4,040	2,586	44
06	4,875	3,120	67
07	3,644	2,332	63
08	3,675	2,352	62
09	3,525	2,256	58
10	2,325	1,488	48
total	188,827	139,606	-
annual average	6,993	5,170	55

[a] the period 1684-1697: exports of Keicho silver (80%)
the period 1697-1710: exports of Genroku silver (64%)

Source: An account book of the private trade, in the So family documents, National Diet Library.

Table 2
The Kind of Silver Cash in Japan and Silver Exports to Korea, during the 17th and the mid-18th centuries

silver cash in Japan[a]			silver exports to Korea		
name	purity (%)	period of circulation	name	purity (%)	period of exports
Keicho silver	80	1601-1738	Keicho silver	80	1611-1697
Genroku silver	64	1695-1722	Genroku silver	64	1697[b]-1711
Hoei silver	50	1706-1722	Special silver (_Ninjindai-okogin_)	80	1712[c]-1715
Eiji silver	40	1710-1722			
Mitsuho silver	32	1710-1722			
Yotsuho silver	20	1711-1722			
Kyoho silver	80	1714-1738	Kyoho silver	80	1716-1738
Genbun silver	46	1736-1827	Special silver (_Ninjindai-okogin_)	80	1739-1750[d]

a Source : H. TAYA, _Kinsei Ginza no Kenkyu_ [A Study of the Silver Mint in the Tokugawa Period], Tokyo, 1963.
b Genroku silver was recognized to export official in 1699.
c,d Special silver was minted from 1710 to 1715, and from 1737 to 1754.

Table 3
The Amount of Silver Currency in Japan and Silver Export to Korea

α amount of issue (index)		β amount of export (index)	α/β x 100
Genroku silver	1,521 ton (100)	71.5 ton (100)	4.7%
Kyoho silver	1,242 ton (81)	minimum 86.4 ton (120)	6.9%

Table 4
The Amount of Monthly Export of Silver Korea: The Seasonal Cycle of Silver Transported

month period	1st	2nd	3rd	4th	5th	6th	7th	8th	9th	10th	11th	12th	total (ton)
1711-20													
T.	1.7	4.3	2.0	2.9	3.8	1.3	2.7	6.5	2.2	5.0	5.8	1.9	40.1
%	4.2	10.7	4.9	7.2	9.4	3.2	6.7	16.2	5.4	12.4	14.4	4.7	
1721-30[a]													
T.	1.4	0.8	0.2	0.3	0.2	0.2	6.0	5.4	4.7	15.2	0.8	0.4	35.6
%	3.9	2.2	0.5	0.8	0.5	0.5	16.8	15.1	13.2	42.6	2.2	1.1	
1731-40													
T.	2.6	2.9	2.1	1.3	0.6	0.4	1.7	5.8	0.6	4.4	6.0	2.6	31.0
%	8.3	9.3	6.7	4.1	1.9	1.2	5.4	18.7	1.9	14.1	19.3	8.3	
1741-50													
T.	0.5	1.4	0.5	0.6	0.6	0.0	1.0	1.1	1.2	1.1	3.0	0.9	11.9
%	4.2	11.7	4.3	5.0	5.0	0.0	8.4	9.2	10.0	9.2	25.2	7.5	
total													
T.	6.2	9.4	4.8	5.1	5.2	1.9	11.4	18.8	8.7	25.7	15.6	5.8	118.6
%	5.2	7.9	4.0	4.3	4.3	1.6	9.6	15.8	7.3	21.6	13.1	4.8	

[a] the period 1721-1730 is excepted the amount of the period 1724-1726 because the documents are not in order.
T. = ton.

Source : Mainikki (diary of the Japanese concession in Pusan), in the So family documents, National Diet Library.

FIGURE 1

Peking
KOREA
Seoul
JAPAN
Pusan
Tsushima
Edo
Kyōto
CHINA
Nagasaki
Satsuma
Fukien
Ryūkyū

FIGURE 2 Tsushima-Korea Trade by Commodities. 1684-1688

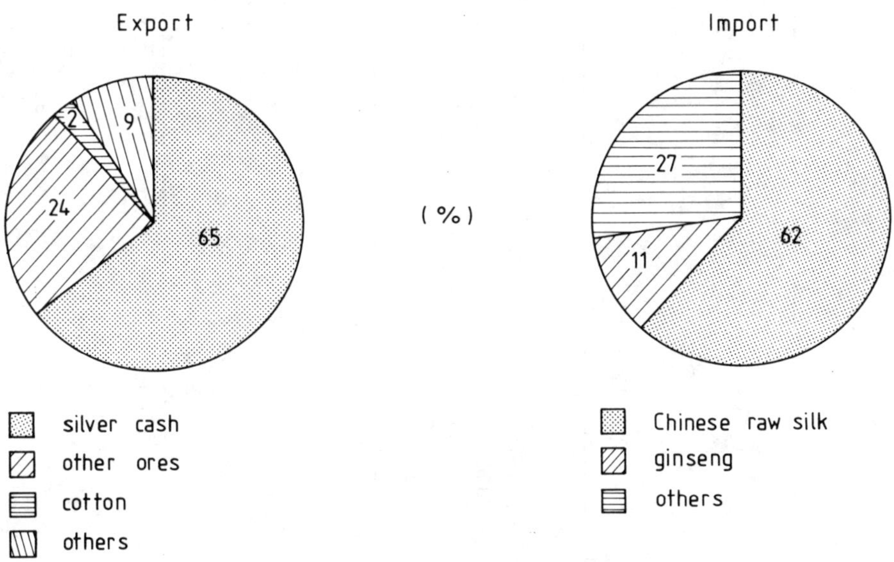

Source: An account book of the private trade, in the So family documents, National Diet Library.

FIGURE 3 : Annual Silver Export from Tsushima to Korea (1684 - 1751).

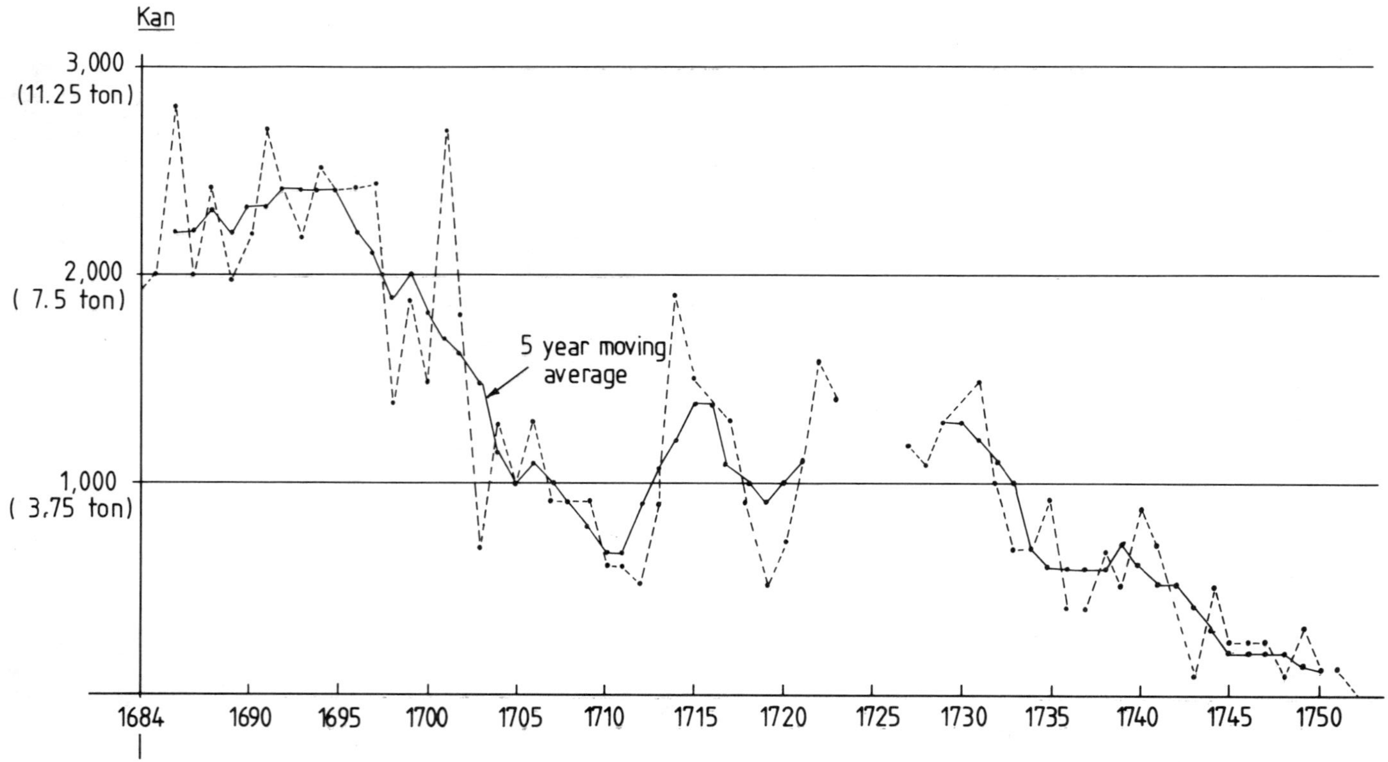

Source: the period 1684-1710 ; An account book of the private trade,
the period 1711 - 1751 ; Mainikki (diary of the Japanese concession in Pusan),
in the So family documents, National Diet Library.

THE PARTS OF THE «MACHINE» DIVISION OF LABOUR IN EUROPEAN AND INDIAN COIN MANUFACTURE BEFORE MECHANISATION

by Frank PERLIN

«'The machine is not the essential feature of the factory.' He found that its essential was its way of organizing the division of labour.»
(F.C. LANE and R.C. MÜLLER, *Money and Banking in Medieval and Renaissance Venice*, Baltimore, 1985, I, p. 234, n. 107, citing K. Bucher).

1. The Problem

The purpose of this short essay is that of pulling together some scattered evidence concerning coin manufacture in 17th and 18th century India, and of interpreting that evidence in the light of what took place elsewhere in the late-medieval and early-modern world. I shall be mainly concerned with a selective comparison of the European and Indian evidence for mint organisation before mechanisation, especially focussing on the more visible parts of the collective work-place. Yet, behind this limited task there lies a more interesting and difficult question concerning the connection between adaptations in the division of labour - in one selected manufacturing activity, and in one region notorious for a relative absence of new, mechanical applications to craft production - and the broader economic changes effecting the early modern world. Elsewhere, I have tried to deal with this question by linking monetary activities in the sub-continent with our growing empirical knowledge of the inter-oceanic maritime trades in monetary raw materials and means-of-payments, a question, nevertheless, which due to paucity of knowledge can at present best be formulated in terms of carefully argued hypotheses and an urgent need to turn more vigorously to the detailed reconstruction of the complex inter-regional patterns of connections of which these trades were part and upon which Indian monetary production had largely depended for supplies.[1] In the present instance, this concern for

[1] F. PERLIN, *Changes in Production and Circulation of Money in 17th and 18th*

connections between local events and wider contexts, has a different, more abstract, if largely implicit emphasis: that of the manner in which a whole complex of technical changes, including tool-differentiation and mechanisation, an appropriate development of numeracy and technical know-how, but also the division of labour, are understood as related constituents in the transition to different, namely capitalistic kinds of economy.

The problem lies in the strong tendency to reserve all aspects of the transition to the European heartlands where a clear scientific culture was in evidence, where a gradual process of selective mechanisation occurred, and where selective industrialisation took place, the last casting a highly problematic, retrospective, and ordering perspective onto interpretation of the past.[2] By contrast, as is well known, non-Western economic histories have the aura of never-could-have-beens, examined for the absence of an Enlightenment spirit and their failure to «progress», and thus for the cultural and political blockages apparently preventing technical or scientific advance. Elvin's formula for China is well-known and characteristic: «quantitative growth and qualitative standstill», and this for a region where, in contrast to India, considerable mechanisation of production and a vigorous scientific culture have been well documented.[3] The result is that hand technology in Asia has

Century India. An Essay of Monetization before Colonial Occupation, in: E. Van Cauwenberghe and F. Irsigler (eds.), Minting, Monetary Circulation and Exchange Rates. Akten des 8th International Economic History Congress, Section C7. Budapest 1982, (Trierer Historische Forschungen 7), Trier, 1984, pp. 323-360. F. Perlin, *Money-Use in Late Pre-Colonial India and the International Trade in Currency Media*, in: J.F. Richards (ed.), Imperial Monetary Systems in Early Modern India, Dehli, 1986. F. Perlin, *Monetary Revolution and Societal Change in the Late Medieval and Early Modern Worlds*, in: Journal of Asian Studies 46 (1986) (Fort Leoming).

[2] A point most recently made in the debate over «proto- industrialisation», for which see Mendels, 1982.

[3] Thus Habermas (J. Habermas, *History and Evolution*, in: Telos 34 (1979), pp. 38-39), representing a hallowed standpoint among «social scientists» («Only [in Europe]...did the innovation potentials of craftsmen and in part also of mercantile elites so consolidate that universalistic principles were abstracted from the world-views and *institutionalized* in the area of social labor»), although more critical where «mentalitees» are concerned; the distinction forms the doubtful basis for a variety of theses claiming some or other allegedly unique property of «Western societies». «Quantitative growth and qualitative standstill» is the title and thesis of the penultimate chapter of M. Elvin, *The Pattern of the Chinese Past. A Social and Economic Interpretation*, Stanford, 1973, but is difficult to reconcile with the previous chapters in which the emphasis is laid upon considerable development in commercial organisation, of large centralized production centres and workshops in crucial sectors such as metalurgy and textiles, extensive domestic outwork, wage-labour, private accumulation of large capital, monetisation and mechanization. For India, see I. Habib, *Potentialities of Capitalistic Development in the Economy of Mughal India*, in: Journal of Economic History 29 (1969), pp. 32-78 and T. Raychaudhuri, *Non-Agricultural Production: Mughal India*, in: T. Raychaudhuri and I. Habib (eds.), The Cambridge Economic History of India, Cambridge, 1982, I, pp. 261-307.

too easily become a category unworthy of investigation, since its basic, epochal, apparently static characteristics can be assumed from one or two relatively well-known or accessable cases. I shall not attempt, here, to justify my defection from the conventional reading of India's essentially hand-based technologies, since the essay speaks for itself. But, it is as well to emphasize that the conventional categorisation of the world described above is now in irreconcilable conflict with increasing knowledge and sensitivity to the more international aspects of change in the early modern world, and with new research - as yet unabsorbed - slowly accumulating on non-western economic history.

Thus my question: if it can be shown that the organisation of non-mechanized manufacture changed in response to new and larger markets for coin in the chosen period, how should we conceive these changes in relation to those more localized developments elsewhere where more of the constituents of the technical packet of changes associated with transition, can, as in Western Europe (and China), be identified? How should we try to characterize the international dimensions of these problems in an age where national histories no longer seem sufficient to understand the course of early modern economic history?[4] While evidence for a limited case such as coining must serve to throw attention onto a host of other features of the commercial economy and to a much larger task of reconstruction in order to answer this question in a satisfactory way, it must also be obvious that even this limited case can stand for itself as a provisional comment upon the possibility of comparing and synthesizing essential aspects of Indian and European economic histories during this period of major change.

2. Combinations of the Production Line

The basic difficulty in dealing with the Indian case lies in the rarity of research into the details of mint organisation and manufacture, while what exists is largely focused on a small handful of single cases, treated as representative. The consequence is that known evidence is extremely scarce, scattered and fragmentary, and not yet brought together in a comparative framework.[5] For example, most of the basic features of coin manufacture

[4] The reference here is to the question of the «unit of analysis»; an international framework composed merely of national histories and «cultures» does not begin to face up to the problem and indeed presupposes it (which is not, of course, to deny national or cultural dimensions to history).

[5] A recent attempt is O. PRAKASH, *Foreign Merchants and Indian Mints in the 17th and Early 18th Centuries*, in: J.F. RICHARDS (ed.), Imperial Monetary Systems in Early Modern India, Delhi, 1986.

have not been studied at all. No studies of die-manufacture or die-productivity exist; crafting and supply of dies have yet to be recognized as problems affecting the quality and supply of coins. The same could be said for many other features: take that of the supply of raw materials to the numerous inland mints of the 17th and 18th centuries, the production of the tools used in minting, or the nature of the skills variously involved in different tasks and in producing varying types of coin, questions about which we know all but nothing. For these reasons, I have tried to compare scattered evidence for mints situated in different parts of the sub-continent during the long period from 1600 to c. 1825.[6] Like much early- and late-medieval European data, such evidence is highly heterogenous and normally presents serious difficulties of interpretation and processing; for the most part we obtain but a single observation for each mint , concerning but a single moment in time (say, one year or part of a year where output is concerned, or the brief moment of a chance visit where observations concerning the work-force are concerned), and this often makes little (or misleading) quantitative sense until set within the particular local conditions prevalent at the time.[7] Yet in spite of these difficulties, the effort is worthwhile: it seems to me that this evidence, when brought together and placed within the comparative perspective of what we know of European production, provides eloquent insight into the adaptive character of changes in mint-organisation during the period concerned.

I shall set the terms of the argument by developing a contrast between, on the one hand, the organisation of skills in the late medieval European and Indian craft workshop, and, on the other, evidence for some moderately important regional mints in India observed in the early 19th century, right at the end of our period. The reason for proceeding thus is to solve gaps presented by lack of evidence for medieval production in the Indian case and to begin with a schematic statement which I shall thereafter qualify and elaborate. First, I shall turn to the significant, if problematic engravings by Burgkmaier and Holinshed of the late-medieval European mint, as depicted

[6] Evidence comes from published and unpublished primary sources, and from a small number of cases described in secondary works.

[7] Mints used or established by the European trading companies form partial exception (Surat, Rajmahal, Murshidabad, for the first; Bombay, Palikat, Madras, Calcutta, for the second), largely because of the interest which European language sources have occasioned. But, as will be seen, the data is paltry and often difficult to interpret, when compared to statistics presented by, say, J. Craig, *The Mint. A History of the London Mint from A.D. 287 to 1948* , Cambridge, 1953; F.C. Spooner, *L'économie mondiale et les frappes monétaires en France (1493–1725)*, Paris, 1956; J.H. Munro, *Bullion Flows and Monetary Contraction in Late-Medieval England and the Low Countries*, in: J.F. Richards (ed.), Precious Metals in the Later Medieval and Early Modern Worlds, Durham N.C., 1983, pp. 97–158. However, even East India Company records have yet to be used in a consistent and systematic manner for questions surrounding the mint.

during the 16th century, that is at the start of the commercial revolution of early modern times.[8] These ostensibly show a handful of craftsmen collected together side by side to pursue the individual processes which together enabled the factoring of coin (in fact its last stages following the refining of the metals). The two illustrations - taken at their face value - suggest a similar distribution of workers per task, a hamerer flattening the pieces of metal, a cutter making the blaks, a striker, an assayer, and in the case of Holinshed a melter. In spite of a different arrangement of working space in the two cases, the principals, if not identical appear relatively similar. To what extent do these engravings illustrate the actual distribution and division of labour in the average late-medieval mint? What characteristics of pre-mechanized mint production, if any, do they represent?

We also possess a classical account of the workmen in the mint of the Mughal imperial household at Agra in the mid- or late-16th century.[9] While no overall description is provided of the organisation of work-space or of the numerical proportions of the different crafts combined in the complete workforce, the account nevertheless provides a sense of a small number of highly skilled, even fine-arts craftsmen, gathered together for what, by later conventions, appears to have been a costly procedure (as we shall see below).

Against these depictions of space and task, we may set an observation of work inside the mint of Chandore at the start of the 19th century.[10] Chandore

[8] Lane and Müller cast doubts on their value, F.C. LANE and R.C. MÜLLER, *Money and Banking in Medieval and Renaissance Venice*, Baltimore, 1985, I, pp. 223-224, n. 67 and I shall comment on this below. Holinshed's engraving may be found on the cover of R.B. OUTHWAITE, *Inflation in Tudor and Early Stuart England*, London, 1969 («depicting the new coinage of 1560-61»), and Burgkmaier's in J. CRAIG, *The Mint*, opp. 42, pl. 1 (a modernized version wrongly attributed to «Coining in Paris c. 1500»); PH. GRIERSON, *Numismatics*, London, 1975, p. 107 (referring it properly to Burgkmaier and the early 16th century); F.C. LANE and R.C. MÜLLER, *Money and Banking*, p. 223 (referring it to an 18th century version of the manuscript); M. DE ROOS, *Sij Ontboode te Coomen Rekenen: Munt en muntpersoneel in Holland in de vijftiende eeuw*, (Projekt Bureaucratisering. Instituut voor Geschiedenis), Groningen, 1978, where it appears on the cover of her essay on mints in 15th century Holland. Lane and Müller's comments are at least timely in reducing the representative character of such illustration.

[9] A. FAZL ALLAMI, *The A'in-i Akbari*, New Delhi, 1977, I, pp. 16-38.

[10] *Nasik Gazetteer*, 1883, p. 429. Indian data generally requires processing or qualification and the reader is referred to the respective tables in F. PERLIN, *Money-Use in Late Pre-Colonial India and the International Trade in Currency Media*, in J. RICHARDS (ed.), Imperial Monetary Systems in Early Modern India, Delhi, 1986, in this case of output Tables 4.1.1-3. With respect to what follows on contemporary description of medieval Indian minting, little credence should be given to the realism of the «Scene in a Mint», a detail of an Indian miniature in the British Library, Department of Oriental Antiquities, *Album of Portraits and Miniatures*, Add 21154 (no. 1974, 6-17-09 [27a]; reference from Harbans Mukhia). I am thus confined to considering Abu'l Fazl Allami's description, which at least treats minting as a complex process subject to a division of labour.

was a small market town in the Deccan uplands of Western India, one of a handful of mints producing a regionally important trading and fiscal coin (the Chandore rupee, circulating in a large part of the Western Deccan), to be contrasted, however, with major mints in the port cities and larger, commercial, urban centres. At the time of observation the region had become subject to severe agrarian crisis affecting both trade and monetary activity. Nevertheless, some 450 workers were collected together: including 400 copper- and gold-smiths to cut the silver into blanks for the coin, and 3 strikers to stamp it; the capacity of the mint was 20,000 silver rupees per ten hour day. In Central India, at the same time, several mints together produced the different kinds of coin required for local and supra local needs. Each mint operated, therefore, on a smaller scale than that at Chandore, producing between 8,000 and 10,000 rupees per day (a little more at the largest mint at Indore): we read of some 50 cutters and 10–12 strikers, with a total work force of 68–72. At Indore, pairs of strikers replaced one another every two hours, implying up to five or six pairs working partly simultaneously and partly in shifts in the production of up to 11,000 coins per day (or 3 to 4 teams, if composed of groups of 3 workers).[11] At Bagalkot, in Southern Maharashtra, where a moderately important regional trading rupee was factored, at a capacity of some 20,000 coins per day, four strikers relieved one another in pairs at the hammer, implying a concentrated production line similar to that at Chandore.[12]

These figures represent some of the fragmentary observations available on the more visible parts of the production line, and derive from the difficult early years of the 19th century; they represent mints supplying the more important precious-metals monies to a broad region in which a much larger number of mints produced a variety of different kinds of coin to serve diverse types of market. They also complement the much more rare information currently available on the mint-workforce during earlier centuries, and a small ammount of suggestive data for productive capacity and actual output of coin from different mints. Thus, early in the career of the then very important Surat mint, in 1647, we read of a daily output of 30,000 rupee-coins, which must have increased in the later 17th century in order to cope with the substantial growth in imports of mintable metal into Surat during this period.[13] Thus, if we assume that all imported silver was coined in the

[11] J. MALCOLM, *A Memoir of Central India Including Malwa and Adjoining Provinces ...*, London, 1823, II, pp. 84–85.

[12] TH. MARSHALL, *Statistical Reports on the Pergunnahs of ... South Maratha Country*, Bombay, 1822, pp. 152–156. From the evidence of working-practices carried out elsewhere (at Bagalkot, for example), and given the arduous character of die-striking, we must in fact assume that there were probably two striking teams at Chandore, alternating with one another, thus 5 to 6 workmen in all (which does not affect estimations of ratios for a working production line).

[13] On the career of the Surat mint, see F. PERLIN, *Money-Use in Late Pre-Colonial India*, Table 4.1.1.

Surat Mint, as officially ordered and as generally supposed to have been the case, we obtain an annual figure for real output in 1640 of about 2.5 million rupee coins (27,000 kgms estimated from all foreign sources [14]), or 10,000 per day, equalling about one third of then current capacity - but a figure much more modest than those given by van Santen.[15] By the end of the century, an estimated 100,000 kgms of silver were being imported into Surat each year[16], which if all minted would approximate to 9 million coins, or 36,000 per day (Indian ships alone were said to be carrying to Surat between 5 and 6 million rupees value of silver per annum in the last three decades of the century).[17] Neither of these figures include the continual recoinage of aging silver coin carried from sub-continental sources and an intrinsic aspect of the Mughal currency system,[18] or, in the 1640s, the probability of silver still being re-

[14] J.J. BRENNIG, *Silver in the Seventeenth Century Surat. Monetary Circulation and the Price Revolution in Mughal India*, in J.F. RICHARDS (ed.), Precious Metals in the Later Medieval and Early Modern Worlds, Durham N.C., 1983, pp. 480-482. J.F. RICHARDS, *Introduction*, in J.F. RICHARDS (ed.), Precious Metals in the Later Medieval and Early Modern Worlds, Durham N.C., 1983, p. 25.

[15] H.W. VAN SANTEN, *De Verenigde Oost-Indische Compagnie in Gujarat en Hindustan (1620-1660)*, Leiden, 1982, pp. 75-76, 84 and Table V.

[16] H.W. VAN SANTEN, ibid.

[17] H.W. VAN SANTEN, *De Verenigde Oost-Indische Compagnie*, pp. 75-76. I use a notional (and slightly low) round figure of 11 gm. per rupee for silver kg. conversions, based upon Bruce et al, 1982, 39-41 and 44-45 (for Surat issues of Shah Jahan and Aurangzeb, respectively). Van Santen's estimates of imports by Indian ships, alone, before 1650 (most likely limited to the 1640s), is between 2 and 4 million rupees value of silver (H.W. VAN SANTEN, ibid.), while, later, the same estimates are attributed to imports by all merchants, see H.W. VAN SANTEN, *De Verenigde Oost-Indische Compagnie*, p. 85. Even a rough idea of the degree to which the authorities succeeded in guiding imported silver into the mint is unclear, information being intermittent, while silver in some form was allowed to be transported inland for the coining of the locally-important South Gujarat mahmudi (thus, one year before the reopening of the Surat mint, a consignment of rials were stolen on their way to Malher for coining into mahmudies. W. FOSTER, *The English Factories in India (1618-1684). A Calendar of Documents in the India Office. British Museum and Public Record Office*, Oxford, 1906-1927, I, p. 91). On the other hand, contemporaries such as Ovington (J. OVINGTON, *A Voyage to Surat in the Year 1689*, ed. H.G. RAWLINSON, Londen, 1929, pp. 131-132) commented upon the absence of forged Mughal coinage, and it may be guessed that a successful «policing» of the rule concerning imports would have made forgery uneconomic since the value of «all» silver would already include a value-increment for mintage. Would this addition of «increment» also help to explain the reason for currently minted coin exchanging at higher value than older coin of the same weight and fineness (a feature of the Mughal currency system)?

[18] E.g. I. HABIB, *The Currency System of the Mughal Empire (1556-1707)*, in: Medieval India Quaterly 4 (1961), pp. 3ff. and K.N. CHAUDHURI, *The Trading World of Asia and the English East India Company 1660-1760*, Cambridge, 1978, p. 183). In 1770 it was said that «silver of different Species to a large Ammount was brought when the Mint flourished from up Country to be coined there and would do so again», but for the fact that proposed duties would «entirely destroy this branch of Mint

leased for minting from stocks of treasure, during these years of recovery and growth. Moreover, these estimates do not take into account what must have been an even larger volume of copper-minting in Surat during the later period; it is an unfortunate fact that all observations, almost without exception, concern only the precious metals, but again, while there were other important uses for the metal, as in armaments, the authorities were recorded exerting pressures on merchants to import copper and convey it directly to the mint, while the Bombay mint reported shortages of copper inhibiting its potentially profitable operations in the 1670s.[19] Thus imports of copper, although low in the 1640s when Mughal copper minting had temporarily all but ceased (in 1645, 16,986 kg. of Japanese copper, which is equivalent to 850,000 dam coins if all were minted), had reached extraordinary quantities in the last decades of the century.[20] In 1684, more than 625,000 kg. of Japanese copper were imported into Surat, being equivalent to 48 million dam coins, or 192,000 per day if all were minted at the local mint. [21] While a peak year, shipments of between half and three-quarters of that quantity were regularly recorded during the previous decade.[22] If only a small proportion of copper supply had, in fact, been minted there would still have been room for spectacular rates of output, implying a considerable workforce and helping to

---

Revenue» and cause silver to be coined «in any one of the out mints rather than here» (Government of Maharashtra State Archives Bombay, *Surat Factory Diary*, no. 657, 1770, p. 52).

[19] For the view that the bulk of copper was minted, see H.W. VAN SANTEN, *De Verenigde Oost-Indische Compagnie*, p. 106, and brief, if unargued statements by H. FURBER, *Rival Empires of Trade in the Orient 1600-1800*, Minneapolis, 1976, p. 256 and, for Portuguese shipments in an earlier period, Vilar, 1976, pp. 97-98. Glamann's evidence for Golconda, on quantities specifically stated as being destined for minting, is revealing, see K. GLAMANN, *The Dutch East India Company's Trade in Japanese Copper (1645-1736)*, in: Scandinavian Economic History Review 1 (1953), p. 50. To imports of Japanese copper, listed above, should be added supplies from Persia in the last decade of the 17th century, and from Europe in the 18th. Shortages of copper for minting at the Bombay mint in the 1670s are recorded by CH. FAWCETT, *The English Factories in India 1670-1684*, Oxford, 1936-1955, I, pp. 75-76, 108-109 and 129, while considerable demand for both the metal and the coin existed in the hinterland at this time, as evidence from Bombay again suggests (CH. FAWCETT, ibid). A letter of 1662, from Surat, cited by W. FOSTER, *The English Factories*, III, p. 77, mentions the current imposition of the rule on the Dutch, while in B.A.J. KEMPERS, *Journaal van Dircq van Adrichem's Hofreis naar den Groot-Mogol Aurangzeb (1662)*, (De Lindschoten Vereniging 45), The Hague, 1941, p. 51 we read the claim that it was actually written into the deed (farman) allowing the Dutch to trade in Surat and elsewhere.

[20] K. GLAMANN, *The Dutch East India Company's Trade*, Table IV, pp. 64-65, Table IV.

[21] K. GLAMANN, ibid. For weights of Surat dam copper coins of Shahjahan and Aurangzeb's reigns, I have rounded out data given in Bruce et al, 38, III, for the 1645 conversion (20 gms.) and 43, II, for the 1684 conversion (13 gms.).

[22] 1684 is the last figure given by Glamann and was probably exceeded in the years thereafter.

explain the employment of the 30 pairs of strikers reported at work in the mint on both copper and silver coining.[23] Finally, putting speculation aside, there was an average real output at the important Murshidabad Mint, in 1722 (or 1728), of c. 48,000 quality coins (Mughal rupees) per day, implying a work-force which, if organized as a single production line like that at Chandore, would, most probably, have reached, even exceeded a thousand (and if organized as at Surat, on which below, a much higher number).[24] There are many difficulties about how to interpret the early European engravings. Lane and Müller[25] accept arguments that Burgkmaier's is essentially an illustration of the crafts of the mint, rather than of a realistic organisation of work space, and thus more of a stylistic statement than a visual observation, thus equivalent, in this sense, to the description of the crafts of the mint at Agra. Indeed, the depiction of what appears to be a crafts workshop hardly fits the occasional evidence for European minting detailing numbers of workers, and relative proportions of the different crafts, although, as we shall see, the manner in which these were combined for work is a less simple matter than at first seems to be the case. At least, in medieval Venice there existed a complex division of labour and work space, production in the second half of the 15th century appearing to have taken place in several «shops» within the mint, among which were distributed the different activities making up the hand-production line; by the 16th or 17th century, 16 of these craft shops were occupied by the sizers and cutters of the blanks (8 in 1425, 12 in 1498), each the focus of a group of specialist workers sharing the furnaces around which the shops were built.[26] Moreover, if judging by the numbers of furnaces built into the mints of London and elsewhere, a similar shop-like structure may have operated in England, thus lending more visual credibility to our 16th century engravings if we assume their focus to

---

[23] J. SPLINTER STAVORINUS, *Voyages to the East-Indies by the Late John Splinter Stavorinus*, tr. and annotated by S. HULL WILCOCKE, London, 1789, II, pp. 475-476, but more than likely to have concerned an earlier period. In the period to which Stavorinus's visit refers (1774/75), the Surat mint was in decline; moreover, his description of current coinage in Surat (J. SPLINTER STAVORINUS, *Voyages to the East-Indies*, III, p. 10) seems, at least in part, to have been taken over from J. OVINGTON, *A Voyage to Surat*, p. 132.

[24] For 1722, K.M. MOHSIN, *A Bengal District in Transition: Murshidabad (1765-1793)*, Dacca, 1973, pp. 10 and 123, n. 2, citing S. BHATTACHARYA, *The East India Company and the Economy of Bengal from 1704 to 1740*, London, 1954, p. 107, who in turn cites J. GRANT, *Analysis of the Finances of Bengal*. For 1728, W.W. HUNTER, *Statistical Account of Bengal*, London, 1876, IX, p. 174, citing BLOCHMANN, *Journal of the Asiatic Society*, I, no. 2, p. 90 (from the «rent role» of that year). The figure for output is calculated from seigneurage income (and thus with a margin of error for possible gold issues).

[25] F.C. LANE and R.C. MÜLLER, *Money and Banking*, I, p. 224, n. 67.

[26] F.C. LANE and R.C. MÜLLER, *Money and Banking*, I, pp. 222-224, n. 67, and 235-236.

have been one of several shops into which the enclosures of the mints were divided.[27] However, while Venice, as a major international trading port, was subject to an unusually intense and varied demand, producing coin for the Mediteranean maritime trades and for «export» to England and the Near East, as well as for a large economic hinterland beyond the city itself[28], it remains obvious that a varying degree of organisational complexity characterized mints elsewhere. Whether we look to the 15th century or to the 18th, it is equally obvious that much depended upon the size, importance and market-functions of the mint, and upon the institutional setting within which mint production occurred. Moreover, even when we look to the affects on organisation and output in Europe of early attempts at mechanisation, we again find the process to have been both selective and highly uneven in its results from mint to mint, it taking a long time before machinery could outpace the production rates of a moderately well-organized, hand-crafting line (certainly, when compared with rates in India). This was true of centralizing systems of minting, as in England, but even more obviously the case in regions where minting remained dispersed, as in the Low Countries.[29] In fact, available figures for output in medieval Europe suggest a mainly leisurely rate of production, broken only by emergency conditions resulting from occasional recoinages, but which latter, not incidentally, were often responsible for the more spectacular production levels occasioning many of the observations on which discussion is based.[30]Thus the recoinage at Durham in 1300 produced coins of lower than average quality at the modest daily rate of 6,500 (given a 7 day week under such emergency conditions; 9,200 per day if we assume the five day week upon which my Indian calculations are based) [Mate, 1969]. The consumption at Bruges in 1468-69[31] suggest a daily production of 13,300 coins, comparable with rates, if we follow Grierson, in French and English mints of the same period; in the late 15th century, Bruges was still a great commercial city and we would expect its mint to have been correspondingly important. At the important Paris mint, during the recoinage of 1640-43, average daily output ranged from a meagre 6,500 in 1640, when issues were almost entirely in gold, up to 19,400 in 1643, by which time silver had come to predominate;[32] moreover, this was accomplished with the

[27] J. Craig, *The Mint*, pp. 41-42 and 56.
[28] F.C. Lane and R.C. Müller, *Money and Banking*, I, pp. 242 and 546-47
[29] See, for example, F.C. Spooner, *On the Road to Industrial Precision: the Case of the Coinage in the Netherlands (1672-1791)*, in: Economisch- en Sociaal-Historisch Jaarboek 43 (1980), pp. 1-18; J. Craig, *The Mint*, pp. 117-118, 123-24, 147, 149-50, 153 and 158ff. Thus, hand dies dating from the 17th and 18th centuries may be found in a number of Belgian and Dutch town museums (for example, the Vleeshuis museum in Antwerp and the Lakenhal in Leiden).
[30] F.C. Lane and R.C. Müller, *Money and Banking*, I, p. 545, for the exceptional 1422-23 figure at Venice.
[31] Ph. Grierson, *Numismatics*, p. 109.
[32] Calculated from F.C. Spooner, *L'économie mondiale*, pp. 483-484 in App. IV; F.C.

help of a newly-installed press.[33] On a second occasion of exceptional output at Paris, average daily output rose from 16,700 coins in 1651 to 25,375 in 1652[34], while in 1733, according to Spooner, designs for a horse drawn press at a mint in the Netherlands predicted a production of 2664 coins per hour (or 26,640 per 10 hour day capacity)[35], both sets of figure being comparable with output in contemporary Surat, and possibly well below. The few figures available for Venice suggest ordinarily low rates of output, but with one precociously high, «probably... exagerated» estimate of over 20 million coins under the probable stimulus of the 1422 recoinage, but also destined for a huge international and hinterland market.[36] At Bologna, a temporary need to replenish the coinage in 1350–51 resulted in an exceptional annual turnover averaging over 5000 coins per day, including 11,616 per diem during one 10 day-stretch.[37] On the other hand, the London Mint apparently produced a remarkable 42.25 million pennies over one 12 month period, during the recoinage of 1279–81, which while suggesting a capacity of some 169,000 coins per day, was most likely accomplished – if Craig's calculation proves correct – with the help of large organized bands of recruits from the Continent, as in 1300 when 263 foreign workers were assigned to London for a further recoinage. Moreover, London was far and away the largest of the 9 «English» mints at work in 1300 (with 30 furnaces, compared to 8 at Canterbury, 4 at Bristol, and 2 at Newcastle).[38]

The highest figure for India yet available is one for an actual annual production of about 12 million coins at Mushidabad in 1722 (or 1728), an average of 48,000/day, which must frequently have been exceeded.[39] But, we need to emphasize one important difference between the European figures and those for production in India, and deriving from the different institutional conditions determining money supply in the two regions: Indian figures refer to «normal» year-by-year conditions, and thus to be contrasted with the recoinages punctuating the course of European minting, which separated decades when much lower and more relaxed output levels generally

---

SPOONER, *The International Economy and Monetary Movements in France (1493-1725)*, Cambridge (Mass.), 1972, pp. 184–185 and 330 in App. A, and 334ff in App. C. In respect of this shift from gold to silver, the figures for output in livres tournois given by Spooner (15 million in 1640, 25 million in 1641, and 20 million in 1642 (F.C. SPOONER, *The International Economy*, pp. 184–185)) are highly deceptive respecting work-line productivity measured in numbers of items.

33 J. CRAIG, *The Mint*, p. 152.

34 Calculated from F.C. SPOONER, *L'économie mondiale*, p. 485; F.C. SPOONER, *The International Economy*, pp. 330 and 339

35 F.C. SPOONER, *On the Road to Industrial Precicion*, p. 8.

36 F.C. LANE and R.C. MÜLLER, *Money and Banking*, I, pp. 545–548 and 549.

37 Salvioni, cited by F.C. LANE and R.C. MÜLLER, *Money and Banking*, I, pp. 538–40.

38 J. CRAIG, *The Mint*, pp. 39–40, 56–57, and App. I.

39 See n. 23, above, for references.

characterized coin factoring.[40] Moreover, there is a further significant difference between Indian mints and those such as at Venice, Paris and London: whereas the latter either dominated national production, as at London in 1279–80 and 1300, and Paris in 1640–3, or serviced a huge international market and a precociously developed hinterland, as at Venice (but also to some extent Surat, where the internal market was concerned), mints such as Murshidabad, Bagalkot and Chandore were only relatively important provincial cases among a large number of mints feeding the same territories.[41]

In terms of kilograms of metal handled by these mints, the highest 15th century figures for output of both silver and gold in all of the Low Countries mints added together ammounts to 10,350 kg. during the decade 1420–9 and 8,528 kg. during 1430–9.[42] Munro also gives a peak rate of 1729 kg. per annum between 1430–34 for the mints of the Burgundian Low Countries, although output never again in the century rose above 1000 kg.[43] Output by weight of silver coinage in the English mints reached roughly 18,000–19,000 kgms in the two exceptional decades of the 1280s and 1300s, and thereafter, until the end of the 15th century, sank to small fractions of these figures (except for about 10,000 kg. in the 1350s).[44] During the recoinage of 1640/3, the Paris mint processed 230,898 kg. in its year of maximum output (which «sums vastly exceeded the combined activity of all the other mints» in France[45]). By contrast, Surat's ordinary *annual* output in the 1640s could have reached 82,500 kgms, or 330 kgms. of silver per day (although, as we have seen, figures for imports, taken by themselves, suggest a certain seasonal concentration at this time); the 1722–28 Murshidabad figure is equivalent to about 132,000 kgms (again assuming a notional 11.00 gms for the 1640s rupee); while that for Chandore at the time of observation could have ammounted to 55,000 kgms. of silver (220 kg. per day), were it not for the bullion shortage which had come to affect production of coin in the subcontinent during these years. All these calculations for India are based on observations of actual daily or annual output, in contrast to the speculative (over-) estimates calulated above from maritime silver imports into Surat, and the figures for recoinages in medieval and early modern Europe. However, as is shown by comparison of the Paris and Surat figures, kilogram totals can be highly deceptive when the major concern is to estimate quantities and rates at which items were produced.

---

[40] See, for example, J.H. Munro, *Bullion Flows*, Table 1, pp. 131–132; F.C. Spooner, *The International Economy*, App. C.

[41] On Paris, F.C. Spooner, *The International Economy*, pp. 184–185. On London, J. Craig, *The Mint*, pp. 39ff and 56.

[42] Decennal totals in J.H. Munro, *Bullion Flows*, pp. 136–139.

[43] F.C. Lane and F.C. Müller, *Money and Banking*, I, p. 548.

[44] J. Craig, *The Mint*, pp. 410–411, App. I.

[45] F.C. Spooner, *The International Economy*, p. 185.

The purpose of retailing these figures is to establish what should now be an obvious point. Lane and Müller describe the Venetian mint as representing «an ancestry...[leading] from family workshops of specialized artisans ... to larger more elaborately equipped workshops». They contrast this schematized evolution with that which leads to «the formation of capitalistic machineless factories» (a distinction to which I shall return more critically below).[46] We may thus compare the workshop structure at Venice with the division of labour observed at the average Central Indian manufactury with its partly simultaneous teams of strikers. At Venice, the ratio of cutters of blanks to strikers is estimated at 2:1[47], while in the Burgundian Low Countries the ratio to strikers of a lump category, «ouvriers» (and including cutters, sizers and refiners), is estimated at 4:1[48]. Interestingly enough, the Holinshed ratio for «ouvriers» and strikers would be 3:1, again lending credibility to his design if we assume it to be focussed on one of several «shops» into which the mint-enclosure had been divided.[49] In Central India, however, the cutter/striker ration comes to 4–5:1 (with the addition of refiners to the first - a kind of «ouvrier» category - to 5–6:1), and at Chandore to 133:1. Given the probability that work in Central India included teams alternating by shift, the ratio, when adjusted for an actually working production line, could rise to between 8 and 10:1. In short, even allowing for the greater ease of cutting the thin European coin, the constitution of the workforce in the European examples seems to imply ordinarily more relaxed rates of work and production arrangements than was the case in India, where, however, our examples point to considerable organisational variation.

For example, the mint located at the port city of Surat (and which we shall later examine in greater detail) suggests an organisation of work very different to that at Chandore and Bagalkot: 30 pairs of strikers (30 specialists each with an assistant) were reported by Stavorinus in the late 18th century; given the marginal importance of the Surat mint at this time, such a number must refer to an earlier period when the Surat mint was one of the major mints of the sub-continent and responsible for coining the regular bullion and specie shipments then entering the port, mentioned above (and when circulation of the Surat rupee became especially preponderant in Hindustan).[50] A descrip-

[46] F.C. Lane and R.C. Müller, *Money and Banking*, I, pp. 234–245.

[47] F.C. Lane and R.C. Müller, *Money and Banking*, I, pp. 532–534.

[48] P. Spufford, *Monetary Organization in the Burgundian Netherlands in the Fifteenth Century*, in: C.N.L. Brooke, B.H.I. Stewart, J.G. Pollard and T.R. Volk (eds.), Studies in Numismatic Method Presented to Philip Grierson, Cambridge, 1983, p. 244.

[49] Although Lane and Müller remark that at Venice the whole combination of tasks was distributed between the «shops» into which the mint was divided, a lesser distribution of labour, involved in the cutting and sizing of the coin, occurred in each of several of these, being centred on the furnaces built into them. F.C. Lane and R.C. Müller, *Money and Banking*, I, p. 224, n. 67.

[50] See n. 23 and text to the note for the reference.

tion of the actual management of coin production at Surat, complements this surprising number of coin strikers.[51] The mint appears to have constituted a space in which different moneyers set up their own production teams alongside one another, in short a parcellized shop-like structure, to some extent comparable to that described at Venice and possibly, in this one respect, to that noted in Central India (whether taking an architectural form or not).[52] In contrast, turning to Chandore and Bagalkot, both capable of producing large numbers of coins, it can hardly be doubted that all the workers were organized in a single production line, focusing finally on a single team of strikers working in shifts. We saw at Chandore that one team was considered capable of stamping the output of 400 cutters, eloquent testimony to a division of labour more akin to the «machineless factory» than to the craft workshop and with which the most appropriate European comparison would appear to be the organized use of space in the mechanizing mint depicted by Diderot.[53] As we shall see, nevertheless, what distinguishes Surat from

[51] A. Das Gupta, *Indian Merchants and the Decline of the Surat c. 1700-1750*, Wiesbaden, 1979, pp. 46-47 n. and A. Das Gupta, *The Merchant and his Money. Glimpses of the Western Indian Ocean in the Early 18th Century*, paper presented to the Conference on the Monetary System of Mughal India, Durham N.C., 1981, p. 8. See also Section III below.

[52] In fact, a much closer and more significant resemblance to the «shop»-like structure of the new 16th century mint at Venice is found in early-19th century Central India, concerning alike the cutting and sizing of the blanks ready for striking (see n. 49). In Central India too, we read of teams, composed in this case of three persons, each member of which was involved in a different sub-task; each team sat around a raised fire-place (compare the furnace-centred division of labour at Venice), and each, as at Venice, was possessed of scales and weights, see J. Malcolm, *A Memoir of Central India Including Malwa and Adjoining Provinces ...*, London, 1823, II, p. 82 n. At Pondicherry in 1737, cutting and sizing is again divided into two distinct activities involving separate workmen (one group of which used scales), but whether combined in pairs or as separate blocks of activity is unclear, see E. Zay, *Histoire monétaire des colonies françaises d'après les documents officiels*, Paris, 1892, pp. 317-318. Descriptions of the English mints in terms of numbers of furnaces is equally suggestive of basically similar subdivisions of tasks within the larger division of labour - that is again around the cutting of the blanks, since involving the largest concentrations of workers around a particular activity (for what else would 30 furnaces at London be required) (J. Craig, *The Mint*, pp. 41 and 56). In contrast, turning again to Chandore around 1800, but also to Surat in the 1770s or earlier, and possibly Bombay, we find that the cutters combined the whole task of smithing the blanks, each therefore (not each team but each «cutter» in whom the «team» was now concentrated) possessing a scales. On the one hand, Chandore, we remember, seems to represent a unified production line (a non-shop structure), while Surat, despite a parcellized production process, nevertheless attained remarkable output levels. *Nasik Gazetteer*, 1883, p. 429; J. Splinter Stavorinus, *Voyages to the East Indies*, II, pp. 475-476; F. Pridmore, *The Coins of the British Commonwealth of Nations to the End of the Reign of George VI 1952*, London, 1975, IV-1). See also n. 112.

[53] Ch.C. Gillespie (ed.), *A Diderot Encyclopedia of Trades and Industry. Manufacturing and Technical Arts in Plates Selected from L'Encyclopédie, ou Dictionnaire*

Chandore is less a question of «rationalisation», so-called, as of conflicting managerial priorities, according to which the massive craft workshop and the single-production-line manufactory were equally appropriately organized.

These observations can be supplemented by returning to the question of the rate of striking per productionteam. On the basis of recent experiments using medieval hand-production techniques, Sellwood has estimated that a team of three would have been capable of striking one coin every 5 seconds.[54] At Venice, the rate is estimated as one coin per 72 seconds in the case of billon piccoli and 4 minutes in the case of silver grossi, figures which must reflect the slack organisation of the production line, and arguably due less to the slow work of the strikers than to the limit placed upon the number of cutters - thus the low cutter/striker ratio.[55] The horse drawn press of 1733, referred to above, was supposed to strike coins every 1.35 secs.[56] The rates at the concentrated hand-production lines at Bagalkot and Chandore are thus remarkable, given the thickness of the Indian coin: strikers were seen striking coins every 1.8 seconds.[57] It is not surprising, therefore, that the death of a single striker at Rajmahal in 1676 resulted in a dramatic decline in the productive capacity of the whole mint.[58]

A look at skill-content confirms these contrasts. In the earlier period mint workers appear to have been few and skilled; thus in medieval Europe we encounter professional groups migrating to places where occasionally exceptional demand arose for their work. Craig describes 450 workers recruited from Paris, Bruges, Brussels, Brabant and Namur for the English recoinage of 1300, over half of which (263) were assigned to London[59] - we have already noted the marked decline in the quality of coins produced on this occasion. Similarly, during a moment of exceptional demands upon output at Ghent in 1433-34, a large number of «foreign» strikers were employed, in numbers which would seem to necessitate a proportional increase in other sections of

---

Raisonné des Sciences, des Arts et des Métiers, of Denis Diderot, New York, 1959, I, Pl. 198-207.

54 D.G. Sellwood, *Medieval Monetary Techniques*, in: British Numismatic Journal 31 (1962), cited F.C. Lane and R.C. Müller, *Money and Banking*, I, pp. 538-539.

55 F.C. Lane and R.C. Müller, *Money and Banking*, I, pp. 535-539.

56 F.C. Spooner, *On the Road to Industrial Precision*, p. 8.

57 Marshall claimed to have seen one hundred rupees struck in three minutes, on the one hand doubting that the pace could have been maintained, and on the other referring to the use of shift work, see Th. Marshall, *Statistical Reports*, p. 155.

58 According to Streynsham Master (in R. Carnac Temple (ed.), *The Diaries of Streynsham Master 1675-1680 and Other Contemporary Papers Relating Thereto*, London, 1911, I, pp. 401-402), «whereas a good Coyner would Stamp in a day 10,000 rupees, we gott not now 700: or 800: rupees, and of these a quater part or more bad Stampt...». Temple, the editor of Master's Diaries, doubts the veracity of the figure of 10,000, but in fact it is those for 700 and 800 which are doubtful, surely an error for 7,000 or 8,000.

59 J. Craig, *The Mint*, pp. 55-56.

the workforce, and thus recruited on a very large scale.[60] Comparable conditions must have occured in India when large tributes had to be paid, during which the production line would have been reorganized and expanded for fast production. 15th century Bahamni Sultanate coppers are thus found in hoards far afield, but these most common issues can also show very marked deteriorations in quality (this was at a time of acute late-medieval bullion shortage, so that copper acted as the major coinage medium). Similarly, the widespread application of a copper-based money-tax on peasant agriculture in Nizam Shahi territories in the Western Deccan, in the early 17th century, is again coterminous with a vast output of shoddy coppers.

Under normal conditions, however, where mint workers included a component of highly skilled craftsmen, labour became a bottleneck factor, only resolved by temporarily lowering the standards of the skills required, the point being that the pre-mechanized production-line could ostensibly be expanded or contracted relatively quickly, but at the cost of quality. The new Bombay mint of the 1670s was beset by labour problems: a shortage of skilled personel, resistance to introduction of machinery (as at Madras, London and Paris in the same period), and a consequent output of gimcrack products.[61] Sokoloff [1984] has recently shown that relatively small additions of manpower and supervision to the old craft family-based workshop could have a marked effect on output. But, given strict quality controls in normal times, the shop-like structure seems to have characterized many mints, some degree of protective corporate organisation among the various kinds of interest concerned with production (including craftsmen), and a relatively developed (if slack and variable) degree of organisation of production in mints even in ordinary times, a ceiling would have been placed on the ability of the mint to respond to the market in a period of expansion and diversification (although the early growth of the mint at Surat demonstrates how flexible even quality workshop production might be).

The description of the personnel of the Imperial mint at Agra before 1600 thus falls into place. Here, we read of persons who can only be described as fine-arts craftsmen: «The Engraver [...] engraves the dies of the coins on steel, and such like metals [...] At this day, Hawla-na Ali Ahmed of Delhi, who has not his equal in any country, cuts different kinds of letters on steel, in such a manner as to equal the [work] of the most skilled caligraphers». As for the cutter, «In Iran and Turan they cannot cut pieces without a proper anvil; but Hindustani workmen cut them without such an instrument, so exactly that there is not the difference of a single hair [...]». While it is clear from the rest

---

[60] P. Spufford, *Mint Organization*, p. 244.

[61] F. Pridmore, *The Coins of the British Commonwealth*, pp. 102 and 156, and several references in Ch. Fawcett, *The English Factories*, I, among which pp. 160–161 and 181–182.

of the account that there were differences in the degrees of skill expected of the different crafts making up mint work, the central purpose was that of the production of a craft product. The considerable rewards paid to the higher grades of worker at Agra confirm this emphasis on skill.[62]

The coins themselves tell a similar story: late-16th century Mughal silver and gold coins are generally (though not always) works of great aesthetic quality: they are characterized by near perfect circles; fine die designs; use of dies whilst in good condition (the same being true, however, for certain earlier coinages, such as 13th century Delhi Sultanate silver coins and 15th century Bahmani silvers and coppers, which display finely wrought, intricate caligraphic designs). The 16th century saw the end of the medieval depression of the later middle ages, and the beginnings of a long-term expansion of the commercial and fiscal economies, stimulating a correspondingly rapid advance in monetisation.[63] At first, this affected the quantity of copper-coin production, but at the end of the century the conversion of the Mughal monetary system from one largely based on copper use and copper accounting to a systematic concentration on mass-produced silver rupees and rupee accounting[64], resulted in a gradual displacement of this «art» object in favour of a more utilitarian product. Cutting accurate circles, if not so obviously an aesthetic pursuit as die-engraving, nevertheless involved considerable skill, and this was especially the case with the Indian coin which was much thicker than its European counterparts. It is in die design and in the relative perfection of the circles, later in the extent to which the striker sought to centre the engraving upon the coin, that compromise made its first regular appearance from the first decades of the 17th century. However, these are changes in quality of crafting which could be introduced without affecting the proverbial high standards and predictability of silver-content and coin-weight of the Mughal rupee.[65]

From the late 17th century, and especially in the 18th, production of coin diversifies in order to serve different kinds of market, a point which can be observed across the whole range of minted metals, but also in terms of the

[62] A. Fazl Allami, *The A'in-i Akbari*, I, pp. 22 and 28.

[63] Magalhaes-Godinho, 1969, pp. 356–357.

[64] I. Habib, *Monetary System and Prices*, in: T. Raychaudhuri and I. Habib (eds.), The Cambridge Economic History of India, Cambridge, 1982, pp. 360–381 and I. Habib, *A System of Trimetallism During the Price Revolution. A Study of the Extent of the Use of Silver-Money in the Mughal Indian Economy*, in J.F. Richards (ed.), Imperial Monetary Systems in Early Modern India, Delhi, 1986.

[65] Concerning the consistent high quality of Mughal rupees, see, for example, J. Ovington, *A Voyage to Surat*, pp. 131–132 and E. Terry, *A Relation of a Voyage to the Easterne India, Observed by Edward Terry, Master of Arts and Student of Christ-Chuch in Oxford*, in: S. Purchas, Hakluytus Posthumus or Purchas his Pilgrimes Contayning a History of the World in Sea Voyages and Land Travells by Englishmen and Others, [s.l.], IX, p. 23.

variety of types of silver coin in the north (all «rupees» of one kind or another) and gold coins in the south (pagodas and their fractions). Diversity takes the form of different standards of workmanship and also of pure-metal content, depending upon the function of the coin.[66] Rather than interpret this merely in terms of a decline in central authority or commercial standards, it is better viewed in more systematic terms. Initially (especially in the later 17th century) there are temporary compromises necessitated by shortages of skilled labour in the context of new coin markets, as at Bombay, while later, towards the end of the period, critical shortages of bullion, resulting from changes in the terms of India's maritime trade towards the end of the 18th century, necessitate similar responses; more generally, however, this diversification of coin qualities and purities should be regarded as numerous separate, yet considered responses scattered through space and time, responding, that is, to the need to produce particular products recognizably distinct from one another, and tending to one side or the other in a continuum of variations between high and low qualities: a trading coinage would need to be both reliable and of reasonable purity, whilst a coin intended for purely local circulation would be less likely to be exported if its quality were recognizably less (an *inverse* practical application of the logic of Gresham's Law, so to speak, but possibly supported and stabilized by the greater cost of producing «inferior», alloyed coin).[67] A range of different territorial scales, types of market and depths of social penetration, characterized the circulation of different coin-types - units of a complex, spontaneously-generated, flexible, but hierarchical system - each type manufactured according to identifiable standards. Fifty or more different types of coin might be found in the bazaar of a particular market town, many of which were of purely local provenance and circulation, although only two, three or four of these were likely to be abundant.[68] While numismatists dispute identities and many remain unidentified, contemporaries - shopkeepers, merchants, the many money dealers found in even the smallest market towns, government treasury officers - seem to have had little difficulty in recognizing and listing them (a know-

---

[66] On this paragraph see F. PERLIN, *Money-Use in Late Pre-Colonial India.*

[67] For argument, F. PERLIN, *Monetary Revolution* for increased costs of factoring small fractions and possibly of lesser qualities, see F.C. LANE and R.C. MÜLLER, *Money and Banking*, I, p. 258, n. 3 (for billon and silver, and for small and large silver coin); and J. CRAIG, *The Mint*, p. 45. At Partabgarh, where the exceptionally poor quality Salim Shahi rupee was factored (for a possibly captive market), seigneurage costs were 2.28% instead of the 1,13 at other Central Indian mints covered by J. MALCOLM, *A Memoir of Central India*, II, 81-84. While I argue lower costs for standards of physical craftsmanship, refining costs may have an inverse relation to pure metal content in cases where quantity of alloy needs to be precise.

[68] Compare B.E. SUPPLE, *Currency and Commerce in the Early Seventeenth Century*, in: Economic History Review 2nd Series 10 (1957), p. 240, n. 1: «In 1614 it was claimed that 400 types circulated in the Low Countries, and 82 in France».

how, however, which, if in a much more rough and ready fashion, would also have been needed by a larger population of coin-users likely to find themselves out of pocket when making errors).[69] Since these coins exchanged against one another, not only in terms of relative purities but those of supply and demand, this ability was a matter of daily practice intrinsic to the currency system.

It is not possible, here, to go into further detail into this development of the currency system, except to remark on its implications for the subject of this essay.[70] It meant that even where silver coin was concerned the skills employed in a mint were closely connected with the type of product required. We can take the earlier Mughal system, with its generally high quality standard weights and purities (sustained even while skills were being reduced for certain tasks), as a notional base line from which to read this change. At the topmost end of the production spectrum, quality coins were still being produced for long distance commerce and government treasuries right up to the end of our period, if subject to increasingly utilitarian crafting. On the one hand, the utilitarian aspect may be measured by the fact of coppersmiths being employed at Chandore and Bombay to cut the blanks for silver trading rupees, a point to be contrasted with the description from the Imperial mint. The converse fact of continued quality control over weights and pure metal content at Chandore is reflected in use by each of the 400 smiths of his own balance and weights for checking the results of his work.[71] In Central India,

[69] Early 19th century East India Company records contain long lists of such coins circulating in different markets, giving their precise exchange values against one another (fluctuating against demand and availability): for example, a series for kasbe Dharwar, in northern Karnataka, and sited on the overlap between the northernsilver and southern-gold currency zones, statements for 1819, 1820 and 1821, in which latter year 42 different silver and gold coins were registered (Government of Maharashtra State Archives Bombay, *Revenue Department*, XIX, 1823, n.p.); a list for kasbe Ahmednagar in 1806, thus before occupation, in which 75 named silver coin-types are mentioned, 11 types of northern gold coin, and 16 of southern gold (India Office Library, *Mackenzie Collection*, General, XIV, VII, pp. 170–74, «Particular Account of the Agriculture...Coins, Weights and Measures of the Tallook of Ahmednuggur...in 1806», Table X). The difference in numbers in the two markets probably reflected the closure of many small mints in the conditions surrounding colonial occupation and bullion shortages. Thus Malcolm's treatment of 8 mints in Central India understates the number and variety existing before 1818; to the south, in the region of Pune, we have definite evidence of numerous small mints in operation, and yet by 1819 the Collector of the new District remarked that the mints in the «city» of Pune were the only ones in «the districts under me», that is the monetary system of the 18th century had already become partly invisible (Pune Archives, *Deccan Commissioner's Files*, Pune Inward Letters, XCVI, n° 555). The implication is that post-occupation lists of coins seriously understate the number previously in circulation.

[70] But see F. Perlin, *Money-Use in Late Pre-Colonial India*.

[71] See also F. Perlin, *Changes in Production and Circulation of Money*, p. 355, Table 5.2, for estimates of variations in weight of this coin. See n. 52.

we typically read that after production the coin was assayed, and that if it fell below standard the whole process would have to be repeated, although this time at the cost of the mint and its staff.[72]

As we descend the ladder of qualities it naturally becomes feasible to employ less skilled personel, thus to tap a much larger reserve of employable labour, first the lesser crafts (thus, it may not be incidental that Chandore was near to an important centre of coppervessel manufacture), and thereafter more humble, but cheaper and more plentiful sources. Here it is worth emphasizing the obvious: lowering the skills employed in the mint was not only feasible when lower-quality standards of crafting had been chosen, but it in turn permitted an increased output per individual worker and per production line, reducing, say, the number of cutters required to feed a team of strikers, or obviating the need for highly skilled die engravers (and the time required to produce a die).[73] The base is reached in a plethora of gimcrack copper coins which flooded late 18th century and early 19th century markets. These might be irregularly shaped, sometimes without any pretense at producing a clearly defined circle or square; the dies would be used to exhaustion and their designs be remarkably shoddy; only a fraction of the die design would appear on the coin, while frequently, in the case of the crudest lozenge shaped coins, only part of the irregular facing on each side would be stamped. Moreover, the range of tolerance within which alloy-content and weight might vary for a particular coin-type, increased dramatically as we encounter this level of copper-gimcrackery, with the corrolary that even such otherwise highly skilled activities as refining and assaying could also be reduced to their most simple dimensions.[74] In short, we enter the most flexible level of

[72] J. MALCOLM, *A Memoir of Central India*, II, p. 83 n.

[73] The differences in times required to engrave dies for quality and baser coinages appears to have been considerable; according to F.C. LANE and R.C. MÜLLER, *Money and Banking*, I, pp. 237–240, a quality die for minting precious-metals coinage could take a few hours, while one of a simpler pattern intended for baser coinage might take about 15 minutes. Under stress, the engraver at Bruges in 1468–69 produced 5.33 per day over an eighteen months period, while in 1492 another Bruges engraver completed over 12 single items per day over a three months period for the Ghent mint alone (a total of 384 pairs) (P. SPUFFORD, *Mint Organization*, pp. 249, 257, nn. 73 and 75). Moreover, in India the steel used for dies intended for silver and gold coining may have been the famous wootz steel of Southern India; steel was observed in use at the Imperial Mint at Agra, at Rajmahal, in the 8 mints of Central India considered by Malcolm, and at Bagalkot, between the late 16th and early 19th centuries; at Bagalkot it was described as «a very hard steel» (respectively, A. FAZL ALLAMI, *The A'in-i Akbari*, I, p. 22; J. MARSHALL, *John Marshall in India. Notes and Observations in Bengal (1668–1672)*, edited by S.A. KHAN, London, 1927, p. 117; J. MALCOLM, *A Memoir of Central India*, II, p. 83 n.; TH. MARSHALL, *Statistical Reports*, p. 155).

[74] On this increasing range, see F. PERLIN, *Changes in Production and Circulation of Money*, p. 355, Table 5.2.

manufacture of all, where numerous small village mints, many of which are today unidentifiable, together with possible use of domestic outwork, were responsible for production of the vast numbers of coins distributed, say, by merchants in advances in some regions to a large and scattered population of peasants and weavers (payments to peasants producing cash crops and to textile manufacturers in village and town), and used at the lowest level of bazaar transactions. The most compelling evidence that what we are witnessing is a sub-continent-wide change in the very principles of mint employment and organisation is provided by data on mint costs. If it is true that there is a decline in the skill content of even high quality rupees, the costs of manufacture, including that of labour, should suffer a corresponding (relative) decline. In a free minting system, from which, by the 18th century, central authority had largely retreated from the internal organisation of many mints and had little to do with many newly-established ones, and where, moreover, circulation patterns often transcended the limits of individual states in the sub-continent, the need to produce a viable coinage in a highly competitive market - together with a frequently explicit purpose of making profits from sales of such coin - would have resulted in costs charged to merchants being reduced close to the minimum (again the exception being where a captive market was concerned).[75] Conversely, it is equally important to emphasize that the need to keep costs low would in turn have set a ceiling on skill-input in a given case. In particular, while a number of factors were likely to result in considerable lack of uniformity in labour costs - among which corporate organisation among craft workers, modes of organisation of the production-line, and the kinds of coin produced, spring especially to mind - this would not have been the case with the seignourage duties levied by sanctioning authorities, in order to secure an income from the mint.[76] The numerical data on costs is difficult to interpret because of its unstandardized nomenclature and content, and because of the many different, often inexplicit ways in which such costs were grouped and quoted. Again, this is not the place to discuss this question in detail, but instead to remark that despite these difficulties - which at present render real costs of manufacture, for a sufficient number of cases, difficult to assemble for comparative purposes - a comparison of seignourage costs between different mints demonstrates a remarkable long-term fall in duties between the 1590s and the 1820s[76a]

[75] See nn. 18 and 66; for effects of the «monetary wars» between Venice and her mainland neighbours c. 1450 in keeping seigneurage low, see F.C. LANE and R.C. MÜLLER, *Money and Banking*, I, p. 258, n. 3.

[76] In fact, the following discussion of seignorage charges represents but a provisional examination of this question. I have yet to attempt a systematic examination of actual factoring costs. The data on seigneurage is tabulated, explained and interpreted in F. PERLIN, *Money-Use in Late Pre-Colonial India*, Table 5.1.5.

[76a] See n. 75.

concerning references and explanations of the following data. At the Imperial Mint at Agra at the end of the 16th century, duties on rupee manufacture ammounted to 5% of output (duties on gold and copper coin were appreciably higher while the actual costs of manufacture were considerable in all cases). At Rajmahal in Bengal in the last decades of the 17th century the duty ammounted to 3.4– 3.9%, and at an unnamed mint in 1714/15 to 2.5%. The last figure is comparable to the duty charged at the nearby mint of Murshidabad in the 1720s and '30s: again 2.5%. At Ahmedabad, in Gujarat, on the opposite side of the sub-continent, one of the cited figures for mint duties prior to 1750 is again 2.5%.[77] When we enter the early 19th century, duties have fallen to very low levels: in Western India, 0.2–0.5% at Chandore; 0.55% at Chakan; 0.5% at Bagalkot, with some slightly higher figures at Pune, where, however, low market demand forced actual duties (0.65–1.25%) well below the official price set by the authorities.[78] Circumstantial evidence for the connection between changes in seigneurage, in the general standards of skills invested in manufacture and in coin functions, and their relationships to market forces, lies in the difference between duties charged on gold coin in the northern silver-based currency zone, on the one hand, and the southern gold-based currency zone, on the other, with their completely different denominational systems.[79] In the silver zone, the Mughal muhr was produced as

[77] That is 2.5% on Muslim sarofs; ostensibly 5% on Hindus; however, given the retreat of Mughal power by the mid-18th century and the fact of Maratha control over much of Gujarat, including Ahmedabad, it is difficult to see the higher rate being charged, let alone collected.

[78] Not all figures accord with the apparent neatness of this series; the charge at Indore c. 1820 was 1.25%, at Nagpur 1823 below 1.0%, while at a small Deccan mint at kasbe Teradal in the 1750s, 0.5% was set and actually collected by the authorities (Government of Maharashtra State Archives Bombay, *Sangli Daftar*, Rumal 232, Bundle 1, papers n° 202990-202991, kaula nama...tankasale kasbe teradala (1753–4), and hiseba tankasala kasbe teradala (1756–7), resp.). Moreover, local conditions seem also to have exerted an influence: at Surat in a late 17th century deed (farman) a rate of 0.25% is mentioned (cited by A. Das Gupta, *The Merchant and His Money*, p. 6 (although a re-translation of the original would be useful)). In the only extant acounts for minting which I have seen, and which appear to concern Surat, no seigneurage charge is mentioned for either copper or silver (F.W. Stapel, *Peter van Dam: Beschrijvinge van de Oostindische Compagnie*, (Rijks Geschiedkundige Publicatiën 83), 's Gravenhage, 1939, II-3, pp. 108–109); interest charges recorded by Das Gupta are entirely concerned with a mechanism to be described below and possibly specific to Surat: «minting time», ammounting to 0.025% per day on the time taken to coin the delivered bullion or specie (A. Das Gupta, *The Merchant and His Money*, pp. 6–7 and 14, nn. 9, 12 and 14; and again F.W. Stapel, ibid.). By 1770, the English East India Company sought to impose a 1.5% duty on their own merchants, and 2.5% on all others, but the competition of neighbouring mints forced the Company to reduce the latter (Government of Maharashtra State Archives Bombay, *Surat Factory Diary*, n° 657, 1770, 51).

[79] See the map in F. Perlin, *Changes in Production and Circulation of Money*, p. 357. On the gold/silver transition see, for example, P.M. Joshi, *Coins Current in the Kingdom of Golconda*, in: Journal of the Numismatic Society of India 5 (1943), pp. 85-95 and J.J. Brennig, *Silver in Seventeenth Century Surat*, pp. 477–496.

a high quality, prestige product, used to some extent in wholesale transactions, but not intended for use on the same scale or for the same utilitarian purposes as the silver rupee. On the other hand, the southern pagoda (together with its fractions) was a normal means of commercial payment, with its lower denominations (especially the fanam) circulating among the general population; in the decentralized political conditions characterizing southern India in the 17th and 18th centuries, pagodas were manufactured in numerous mints in the same way as were 18th century silvers further north. Not surprisingly, then, the technology employed in the manufacture of the southern pagodas was clearly more simple than that used in crafting the northern muhr: for example, in place of the elaborate caligraphy on the Mughal die, pagodas were stamped with simple designs, often on only one side, a procedure reducing considerably the time spent on die-manufacture and easing the tasks of striking.[80] In turn, costs were substantially different: for the mohur, 5.8% at the Imperial Mint at Agra, before the end of the 16th century, and 3.38% at Rajmahal in 1692, and for the pagoda, 0.75% at Palikat in the 1640s and 0.5% at Fort St. George (Madras) in the 1660s and 1670s.[81] Two corrolaries of these various developments need to be spelled out. The first concerns the difficulty in interpreting the small ammount of information available on the division of labour in different mints, since this is likely to have been closely connected with the kinds and degrees of skills required to produce a particular product, as well as with the quantity and rate at which output was ordinarily required. Gimcrack coppers, intended as advance payments to peasant cultivators and manufacturers, would presumably have been paid out at very irregular intervals when merchants made their purchases or gave out orders; trading coinages would arguably have had a more continuous use, while those circulating in small country markets would have fluctuated according to a variety of stimuli, including the intensity of the agricultural year; these differences would have had corresponding consequences for the organisation of work. In fact, other influences - such as generally unknown institutional constraints operating at different mints (and already partly hinted at) - might also affect the organisation of the labour process and I shall be returning to this problem below.

---

[80] These comments are partly based upon examination of part of the collection of pagodas kept in the British Museum Department of Coins and Medals, and I thank Nicholas Lowick and other members of the staff for their advice. It also seems that the blanks of the pagoda might have been cast, both from the shape of the coin, its texture, and the fact that the edges do not appear cut and hammered. See n. 72 on speeds of die manufacture.

[81] However, it must be added that the breakdown of costs at Palikat, as interpreted by O. PRAKASH, *Foreign Merchants and Indian Mints in the 17th and Early 18th Centuries*, in J.F. RICHARDS (ed.), Imperial Monetary Systems in Early Modern India, Delhi, 1986, and compared with those at Rajmahal in the northern zone, do not accord with this hypothesis.

The second corrolary is that decline in quality does not indicate a retreat from sensibly organized commercial practice, say the non-market end of some dual economy; rather, it was coincident with the increasing commercialisation associated, say, with the huge extension of textile manufacture for export, with increased use of coin in wage payments, with the growth of a whole hierarchy of markets in which coins were used for a variety of different kinds of transaction, and thus with regular, mass production of the products of the mint. In short, whether it is a question of the huge coin manufactury at Chandore or of small village workshops in the cotton-growing and weaving districts of Central India, we seem to be measuring uses and divisions of labour which had departed far from the old craft standards of the earlier period (both real and notional).

As Malcolm commented in the early 19th century: «The process of coining in the mints of Central India is more correct and expeditious than could be conceived from an inspection of the mode they adopt; but the division of labour is well managed, and the care and habitual exactness of the workmen supply the place of many of those improvements which European skill has introduced into these establishments».[82] This is the 19th century Englishman's reaction to that same simplicity of technology to which the 16th century chronicler had reacted at the Imperial Mint at Agra; satisfaction in both cases but according to entirely different criteria of comparison.

3. Divisions of Time and Labour

In spite of the different character of currency systems in Europe and India, the broader continuum of forces affecting minting seems to have been similar: larger and more differentiated markets for coin resulting in production of a range of different products, although under differing institutional conditions. Most notable is the increase in circulation among the more anonymous levels of the population, and involving a huge volume of production of low value coinages, for example of the copper monies and tokens which filled the purses of country peasants, domestic manufacturers, urban artisans and labourers in different parts of Europe. Clipping and forging, and the manufacture of tokens, filled the large gaps left by the inability of official mints - subject to the inflexibilities of centrally controlled European currency systems - to satisfy 17th and 18th century appetites for coin. These more informal levels of production involved new kinds and locations of work, changes in skill input and in the organisation of labour, and complemented,

[82] J. MALCOLM, *A Memoir of Central India*, II, pp. 81-83.

on the European side, the increasing informality of the whole structure of legal minting in the Indian sub-continent, where merchants and bankers seem increasingly to have controlled access and organization, and even to have set up their own establishments to supply markets in which they were directly involved (and in respect of which, the European East India Company mints were but examples).

The evidence, although small in quantity, is unambiguous concerning the different combinations of craft skill involved in Indian mints, implying considerable flexibility in local response to new structures of demand. At Chandore and other late 18th and early 19th century mints, we can identify a single production team, while at Surat the evidence suggests an enclosure within which several production teams worked side by side, more than required by a simple need to increase the output of a Chandore-type unit.[83] However, mint organisation involved more than mere crafting of refined metal into coin, upon which I have largely focused above: supervision, management (including bookkeeping, and thus a body of clerks), refining and melting, and the assaying of the final product, involved organisational procedures and interests which allowed considerable further variation, and this variation even affected the point at which the merchant or banker, bearing specie or bullion into the mint, articulated with the autonomous organisation of the mint itself (a feature also registered at Venice).[84] In part, this is a question of the persons and institutions effectively directing mint-management and controlling access, thus of the nature of the coins produced and of the interests involved; at some mints, where a group of merchants or bankers had obtained monpopoly control, or which they had actually founded, no free access existed at all, the whole mint-process ostensibly handled by persons concerned with sale or distribution of the product, a feature which, moreover, was aggravated by the tendency for authorities to farm out actual running of mints to entrepreneurs, as indeed happened at the East India Company mint at Bombay, among many other places.[85] At the important mint of Murshidabad, in Bengal, which produced a trading coin that circulated through much of the sub-continent, access was reported to be ostensibly free to Company traders in 1716 (not a stable feature of the Murshidabad mint, however): it was «Agreed they may Refine and Coin what silver they have in Murshidabad Mint»[86]; at the mint at Kasimbazar, however, we read

[83] Moreover, O. PRAKASH, *Foreign Merchants*, rightly remarks that the procedures of minting discussed by him at late-17th century Rajmahal had different features to those described at the Imperial Mint at Agra; in fact many other differences between mints could be mentioned, for example with respect to refining and to the labour force involved in it.

[84] R.C. LANE and F.C. MÜLLER, *Money and Banking*, I, p. 222.

[85] K.N. CHAUDHURI, *The Trading World of Asia*, pp. 188–189.

[86] C.R. WILSON, *The Early Annals of the English in Bengal. Being the Bengal Public Consultations for the First Half of the Eighteenth Century*, Calcutta, 1900, II, p. 246, n° 979.

of «Futtichund [in 1704] having the entire use of the mint [so that] no other shroff [banker] dare buy an ounce of silver...».[87] Similarly, the very different mint at Partabgarh in Central India was founded and monopolized by «four mercantile houses: no money can be coined but what comes through their hands».[88] But while evidence for other mints show similar, if variable constraints, many cases tend towards a greater degree of «free» access to the mint. It is in these latter that the organisational complications are particular difficult to unravel, because of the opacity and rarity of source descriptions. However, the attempt is worth making since these complexities bear upon the further argument of the paper.

In several Central Indian mints at the moment of observation, it was claimed that «the right of coining is vested in no particular body, or individuals; any banker or merchant sufficiently conversant in the business, has merely to make application to Government [...] engaging to produce coin at the regulated standard, and to pay the proper fees on its being assayed and permitted to pass current».[89] That this may involve the merchant actually «entering» the mint and becoming in some part involved in some of the initial processes of minting is suggested by further, if somewhat ambiguous descriptions of the merchant «having collected a sufficient number of silversmiths»[90], but also by the distribution of costs and payments. Costs are detailed for each of the functions involved in the mint-process; while some - notably refining and casting - are listed as a direct responsibility of the merchant himself,[91] others, especially those for engraving the dies, and for cutting and striking, were paid by the authorities out of the «seigneurage» charge, or else, where the mint was farmed out to, or established by private interests, taken from the charges of manufacture levied by the management in addition to seigneurage (which variations make calculation of a true seigneurage so problematic elsewhere).[92] This apparently loose articulation within the actual compound of the mint may help partly to explain the similarly-disaggregated character of the actual manufacturing-process in Central Indian mints (thus correlating with the low cutter/striker ratio relative to that at Chandore). We have already seen that the Surat mint had an

[87] C.R. Wilson, *The Early Annals of the English in Bengal. Being the Bengal Public Consultations for the First Half of the Eighteenth Century*, Calcutta, 1917, III, p. 377, and see also pp. 369 and 382.

[88] J. Malcolm, *A Memoir of Central India*, II, pp. 80-81 and 84-85.

[89] J. Malcolm, *A Memoir of Central India*, II, pp. 80-81.

[90] J. Malcolm, ibid.

[91] Lane and Müller mention the self-interest of the merchant in the fate of the refining of his supply, and not least over the dross from which some recovery could be made. Thus, see also Ch. Fawcett, *The English Factories*, I, p. 181 for misappropriation by the coiners of the dross left in minting rupees at Bombay, in 1677 (which was said to have gained them 0.6%-0.7%).

[92] See also F.C. Lane and R.C. Müller, *Money and Banking*, I, p. 231.

even looser structure, in which several production lines worked side by side, although the reasons behind this parcellisation of the production-process appear to have differed in the two cases.

In the Surat case, this reason lies less in the conservative character of craft manufacture, as such, than in the manner in which production was managed, which thereby served to foster continuance of a craft-like organisation. In 1661-62, some 300 local bankers had purchased rights to mint coins, thus to set up their teams of workers and equipment.[93] It is to these bankers that the merchant with specie or bullion had to turn, so that in contrast to Indore, in Central India, he was unlikely himself to enter the confines of the mint and was certainly not involved in any of the processes of factoring coin. This managerial structure meant that the authorities had only a formal say in organisation, enforcing standards of weight and purity, and their rule that imported precious metals, at least, and sometimes copper, should be born to the mint.

On the one hand, we can already see that mint organisation adapted itself in different ways to varying local conditions, to the nature of the product, the regularity and intensity of demand (low demand probably reducing the cutter/striker ratio and inducing intermittent, leisurely working habits), and, most important, the different, often conflicting interests of the various parties involved in it, from authorities, to monopolizing commercial interests and money dealers, to the craftsmen themselves. In this respect, it is worth emphasizing that the question needs first of all to be investigated in terms of the complexity of contextual conditions, which thus requires detailed reconstruction so that different cases may be properly compared and contrasted, and changes over time be properly guaged, before concluding that what we are witnessing can be satisfactorily characterized as a linear «rationalisation» of craft production into «conveyor-belt»-type processes (or the absence of such a process). I shall try to demonstrate this point by entering further into the peculiar organisation of the mint at Surat, where private and official interests combined with growing market demand to stimulate a highly special, but effective set of organisational arrangements.

---

[93] For the following discussion I have used the accounts in H.W. VAN SANTEN, *De Verenigde Oost-Indische Compagnie*, p. 80 (the 300 purchasers of coining rights encountered in 1661-2, and viewed in terms of daily rotation); A. DAS GUPTA, *Indian Merchants*, p. 46 and A. DAS GUPTA, *The Merchant and His Money*, p. 8 (undated and unascribed, but the purchasers are seen to be responsible for different production lines working simultaneously); J. SPLINTER STAVORINUS, *Voyages to the East-Indies*, II, pp. 475-476 (the thirty pairs of strikers); K.N. CHAUDHURI, *The Trading World of Asia*, pp. 184-185; F.W. STAPEL, *Peter van Dam*, II-3, pp. 108-109. There is implicit disagreement over the role of the authorities, the form of control and activity exerted by the right holding «tanksal sarofs», and on other details. Evidence suggests that at times the authorities had more control than van Santen and Das Gupta allow.

First, it must be noted that the following discussion depends upon a handful of descriptions in documents and secondary sources of considerable ambiguity and frustrating brevity, distributed over a period of some 150 years (where dates and sources are given).[94] It is hardly likely that mint-organisation remained stable over the whole course of the 17th and 18th centuries, given the radical changes affecting the economy and political control of the port, and the changing fortunes of imports of monetary metals - to assume so would indeed contradict the very tenor of this essay. Yet, these fragmentary descriptions represent much more information than available for most other mints, and while sensitive to the dangers of trying to construct a coherent pattern out of them, they nevertheless portray sufficient organisational idiosyncrasies, for the period when imports and minting were at their peak, to bear the weight of the argument of the text, that is of a handful of propositions drawing sustenance from the basic outlines presented by the sources.

The main problem for those whose livelihoods derived directly from the work of the mint lay in the discontinuous, seasonal character of maritime imports caused by monsoons and trade winds. The result was that the mint too was subject to an irregular supply of monetary media, including periods of the year when apparently no raw materials from overseas were delivered (or, at least, were much reduced). The response to these conditions lay in a mechanism for turning these irregular supplies into a relatively continuous flow of work («discontinuous» into «continuous motion», so to speak), which, *ideally*, would be distributed over the course of the year.[95] This has been called «minting time», whereby merchants bearing bullion or specie to be coined entered into contracts with those who had purchased the rights to mint coin, to get delivered minted coin after a stipulated and often prolonged period of waiting.

In practice, however, merchants generally required more immediate supplies of money in order to make purchases, redeem accumulated bills, and repay debts. Given sufficient supplies of coin in the port, arrangements existed whereby delivery of coin could be obtained more quickly, upon payment of an interest charge covering the nominal period of waiting (after which the actual bullion or specie delivered by the merchant was coined and supplied to others in the queue).[96] Use of the extensive credit supplies of the port was another means of temporarily circuiting the problem, but in the main it must be obvious that the apparent dominance of maritime imports

---

[94] See n. 83, above.

[95] Which is to say that except in the first decades of the 17th century, when the mint opened (1620), when it subsequently became subject to the visissitudes of the 1630s famine, and when increases in imports had not yet achieved their later momentum, maintained capacity was soon outstripped by both supply of «raw» materials and demand for coin.

[96] See n. 75 on «minting time».

over subcontinental supplies of silver into the mint (apparent, because recoining of internal monies is invisible), meant that most customers - once the mint was organized in a manner which distributed supplies over a prolonged period - had often to adjust to a period of waiting before freshly minted rupees were made available. «Minting time» might extend from less than a month to several months: the few cases cited in the literature, stem from the early 18th century and range between 71 and 135 days, charged consistently at 0.025% interest per day[97]; while disparaged by European merchants, it can be interpreted as an organizational mechanism designed to adjust production to the limiting conditions affecting imports of monetary metals and craft-minting, and more particularly to protect the parties employed within the mint from too direct a response by managements and merchants to these contextual pressures.

Two such groups spring immediately to mind, first the approximately 300 money dealers (tanksal sarofs) who had purchased the rights to coin, and second the employees of the mint themselves, especially the craftsmen, who elsewhere in the 17th century had shown their capacity to frustrate attempts to modify production techniques and working conditions. On the other hand, it is important to emphasize that the authorities were observed at least once vainly attempting to regain control from the right-holding money dealers, and for obvious reasons: as the local governor of Surat informed the Mughal emperor in 1661, it would be possible, thereby, to add 100,000 rupees to the income of the treasury.[98]

Take first the position of mint employees directly involved in the physical crafting of the coin (that is as distinct from supervisory and managerial staff). By responding directly and immediately to successive gluts and shortages in supplies of bullion and specie, an enormous reserve labour force would have been required, much of whose time would have been spent unemployed. To the extent that the skills used in factoring the high quality Mughal rupee were available in such increased numbers - and, despite the shortages of the period, the early career of the Surat mint displays a remarkable growth - it was clearly in the interests of craftsmen paid by the piece to limit such growth, and to sponsor some kind of gearing system allowing a relatively smaller number of craftsmen to obtain as continuous an employment as feasible.

Nevertheless, it may also be argued that relative shortages of critical skills (argued above) would have necessitated such a system; but, in fact, the

---

97 A. Das Gupta, *The Merchant and His Money*, pp. 6-7, 14 nn. 9, 10 and 12.

98 H.W. Van Santen, *De Verenigde Oost-Indische Compagnie*, p. 80. At the very low rate of 0.25% seigneurage apparently mentioned in the farman (n. 33), and referred to in n. 33, some 30 years later, this would come to 40 million coins; however, a rate set in the 1660s would most likely have been higher, and the number of coins correspondingly less.

question of copper implies otherwise. While explicit evidence that copper-coining was also subject to «minting time» is inconclusive, at present, the huge seasonal influxes of the metal entering the port in the late 17th century (far outstripping the volumes of silver in the same years) imply such delays, whether consciously organized or de facto.[99] In the later 17th century, when imports were increasing to unprecedented levels, Mughal copper came to be coined by techniques often as gimcrack as anything produced by the Bombay mint during the same period, so that a significant, even dominant portion of mint production could have become open to much more flexible organisational and recruiting procedures than in the case of the precious metals.[99a] In this respect, therefore, the most vulnerable section of the workforce, the craftsmen, would have benefited from the protective screen provided by «minting time».

The position of the second group, the «tanksal-sarofs» (or right- holding coiners), is more problematical; it is clear - both from their purchase of rights, and from concern of authorities to regain control - that minting Surat coin could be highly profitable, although in which aspects, in the case of Surat, is not wholly clear. On the face of it, this group would reap the potential profit arising from the difference in purchase price of the raw material and the face value of the finished product, and they also ostensibly benefited from the interest charged on avoidance of «minting time». It is the latter which might give to the coiners a vested interest in controlling output; otherwise the question for the sarofs is one of the relative benefits obtained from distributing their separate management packets (ostensible rights to one day's coining per year - thus, to continue our metaphor, the conversion of bulk time and quantity into discrete packets) successively over the course of the year, or separately, but simultaneously within the space of the mint enclosure in the form of separate production units - it is plain that both took place. In short, the interest of the «tanksol-sarofs» lay in parcellized production within the mint, a second gearing system reflecting the sharing of control among a large but limited body of professional specialists, and strongly affecting the actual work-process (and hindering development of a more unified type of production line such as that found at Chandore and Bagalkot).

The two mechanisms - «minting time» (conversion of discontinuous into continuous motion) and what we might call «dealers» time (division of the time thus created into discrete packets) - were closely connected with the needs of a busy port, with its large number of bankers and money dealers, the

[99] But see F.W. STAPEL, *Peter van Dam*, II-3, pp. 108-109, for one 17th century case, where one month at the same interest charge of 0.025% per day is mentioned: «for the time in which men are busy in coining the copper and exchanging the 'peysen'».
[99a] For example, BRUCE, et al., p. 43.

seasonal character of its maritime trade and the importance of the Surat rupee as an administrative and trading coin within Hindustan. They were also closely connected, even dependent upon the strong official pressures exerted (at least intermittently) upon importers to mint imported raw materials. In later times, when merchants were able to chose the mint at which supplies were coined, any prolongued period of waiting would have been self-defeating. Thus in the 1760s, differential costs motivated merchants to obtain a somewhat inferior coin at Broach, instead of at Surat.[100] Given the lack of explicit information directly expressing the collective interests of the various groups involved in the mint, further speculation should be resisted. My point, instead, has been to show the presence of a problem needing interpretation, because by treating the mint as a sensibly conceived and defended social arrangement - the typical cumulative result of a series of institutional compromises between different groups (and assuming thereafter a logic of its own, as much determining local money supply as determined by it) - we avail ourselves of a much more open approach to the subject than hitherto. On the one hand, it is clear that response to new markets could take different forms according to the bundle of specific conditions influencing organisation in each separate case; a history of collective work in India and Europe might be less concerned with establishing representative types, or genealogies of representative organisational stages, so much as the manner in which a variety of often idiosyncratic, institutional solutions served to mediate the complex of general forces and specific interests embodied in mint-production. On the other hand, an open, synthesizing framework of this kind - within which a comparative history of early modern forms of work may be reconstituted - enables the mint to be set within a broader and more international packet of transitional events.

In the Surat case, we find mechanisms devised - whether to protect the interests of the sarofs or those of the craft workers - to prevent the kinds of «conveyor-belt» response which would form the stuff of both non-mechanized and mechanized factory production elsewhere. At the same time, however, the result was a mediatory set of arrangements for converting a complex of uneven, often violent contextual forces into a variously composed, production routine, at once preserving pre-industrial autonomies and, by doing so, representing the kinds of creative urge seen to have characterized development of the «mechanical mind».

---

[100] See n. 18 (and p. 21 of same reference).

4. Conclusion

Those who write on the late medieval and early modern European mint, frequently speculate on the extent to which it was an early forerunner of some of the organisational hallmarks of the industrial revolution. Grierson has thus referred to it as «the one true example of the mass production of a manufactured object...before the industrial revolution»[101], and Spufford, in his treatment of late-medieval examples of European coin output, uses a similar analogy: «in order to produce coins at a rate of up to a million a month, a late medieval mint had to be organized as an efficient factory», to which he adds that only the conveyor belt seemed to be missing.[102] With more circumspection, Lane and Müller attempt to place the organized workshop of the late medieval mint into a genealogy stretching from guild regulation of the crafts to a greater degree of centralized supervison, and they show how some supervisory and working practices were either taken over from the guild in the course of the history of the mint or replaced by more production line related tasks.[103] Essentially, however, they oppose this genealogy to the other which would lead «naturally», as it were, into the forms of work deemed characteristic of the industrial revolution. Thus, in interpreting their argument, Pullan emphasizes the essentially pre-industrial craft-type structure which continued to characterize the mint: Venice was «another important example of a communal workshop where craftsmen ... remained craftsmen, and never became workers in an assembly line».[104] The mint is no «proto-factory», in such a view. In fact, Spufford further remarks - thereby reflecting the uncertainty which characterizes this whole issue - that seen in general terms the basic organisation of the mint barely changed after the 13th–14th centuries, when the main formative, organisational inovations were accomplished: there is thus little to distinguish the mint of late 14th century Flanders, those of the Burgundy Low Countries during the 15th century, or that of late 16th century Antwerp.[105] But what is it that has not changed: an indigestable, craft-shop structure and an apparently non-profit oriented kind of supervision, or proto-factory tendencies which have reached

---

[101] PH. GRIERSON, *Note on Stamping of Coins and Other Objects*, in: CH. SINGER, E.J. HOLMYARD, A.R. HALL and T.I. WILLIAMS (eds.), A History of Technology, Oxford, 1956, II, p. 485.

[102] P. SPUFFORD, *Mint Organization*, p. 242; see also I. HABIB, *Potentialities of Capitalistic Development*, p. 69, respecting Surat.

[103] F.C. LANE and R.C. MÜLLER, *Money and Banking*, I, pp. 233ff.

[104] B.S. PULLAN, *Between Silver and Gold*, in: Times Literary Supplement 4320 (1986), p. 67.

[105] P. SPUFFORD, *Mint Organization*, p. 239.

the limits of their potential development within a given form of society? How much credibility should we accord Lane and Müller's dichotomous classification of management/production forms and of evolutionary trends?

Despite apparent similarities between different mints in different periods, and despite the differing emphases on their interpretive significance favoured by different authors, many essential questions remain unsettled, especially as regards the actual combinations and identities of workers involved, in both normal and exceptional times, in an actually functioning production line. In this respect the numerical figures are often as ambiguous and puzzling when compared with one another as the generalisations which issue from them, and they are especially so when contrasted with those from the busy, market-oriented, but variously organized coin-manufacturies of early-modern India. Ratios of cutters of blanks to strikers of the finished coin of 2:1 in medieval Venice, and 133:1 at Chandore around 1800, obviously betoken much more than differences in types of product put out by mints in the two regions (say the differences in the thicknesses of coin conventional to the latter): what seems to be indicated, instead, are different kinds of working practice, whatever the formal resemblances apparently linking different, machineless, collective workplaces. Contrasts between rates of striking at different mints suggest identical problems, both within the different regions (Europe and India) and between them; these may again be partly related to types of coin, but also to more fundamental organisational questions. Nevertheless, there is no simple solution to the problem, say that of the difference between disaggregate working practices at Venice and the single production line of modest Chandore and Bagalkot. For - and here we turn back somewhat ironically to Spufford's formula - the mint at Surat (the most appropriate equivalent to Venice in terms of both market orientation and scale) was far from being organized like a factory (with only the «conveyor belt...missing»): despite parcellized management and production practices, it still managed to produce «up to a million [coins] a month», and possibly much more by the last decades of the 17th century. Venice proves a deceptively simple example.

As remarked, much difficulty, then, lies in interpreting, and thus quantifying the occasional and scattered references to members of the production line, thus of calculating numbers of those actually involved in factoring a certain quantity of coin, and of reconstructing the forms in which they were actively combined. It is often unclear whether such numbers refer to masters alone, or also to apprentices, assistants or even migratory gangs, while the often hidden flexibility of conventional occupational terminology and categories makes it difficult to know exactly which specific crafts were included in a particular documentary reference, again reflecting upon the question of numbers and proportions.[106] For example, European ratios might need to be

[106] Several critical examples are mentioned in F.C. LANE and R.C. MÜLLER, *Money and Banking*, I, ch. 12 and App. B and P. Spufford, *Mint Organization*.

reconstituted (increased) to include a division of striking teams into shifts; shift work was part of the normal routine of a busy Indian mint, it being said that a striking team needed to be replaced after one or two hours of brisk work. An answer to this question might indicate considerable contrasts in organisation between the ordinary slack routine of the mint employing a single team and the emergency conditions attendant upon recoinages, where shift-work might be introduced; the former may be characterized by a disagregate craft organisation, the latter by vigorously «rationalized» production lines constituted by bands of specialized migrants. An absence of shift arrangements, would explain the remarkably slow rates of striking adduced in some European cases, and thus a certain complacency with slow rates of output by early machines. Questions also arise concerning the proportion of listed employees regularly at work in particular mints, or the extent to which such a list merely constituted a core around which casual craftsmen were hired and layed off as circumstance demanded.[107] In short, the very empirical basis on which generalisation is based seems to be betrayed by the scale of the obscurities and ambiguities underlying speculations and calculations for individual cases, and measured by the frequent tendency to fill gaps in local evidence with borrowed descriptions from treatments of other mints in other regions, themselves of an often equally speculative character.

We have seen, however, that observations and figures from the Indian mints demonstrate a variety of organisational arrangements affecting both ownership and management, on the one hand, and the production-line, on the other. In this respect, Lane and Müller make what at first sight seems a useful and common-sensical contrast between «communal and princely centralized workshops ... created mostly in order to ensure safeguarding and the high quality of the materials used and the products» (a category including the mint), and those organized by «private entrepreneurs [who] enlarged their workshops to cut costs, and in so doing...changed more or less the process of production»[108]; we saw that this difference underlay the strong position taken by Pullan[109] against viewing the mint as a proto-factory. This classification is closely linked to their distinction between two kinds of evolutionary process affecting collective work, also noted above, the one leading to «larger, more elaborately equipped workshops», the other to «capitalistic machineless factories». The question, however, is whether or not this type of division (and the assumptions upon which it is based) does justice to the varied evidence

[107] Thus J. Malcolm, *A Memoir of Central India*, II, p. 83 n., remarks on the possibility of employing extra refiners in case of need, using their own fluxes, and clearly refers to a fund of specialists outside the mint; F.C. Lane and R.C. Müller, *Money and Banking*, I, pp. 220 and 534, refer to the contracting out of work to refiners in the city and possibly of other tasks to domestic outworkers.

[108] F.C. Lane and R.C. Müller, *Money and Banking*, I, p. 235.

[109] B.S. Pullan, *Between Silver and Gold*, p. 67.

for different organisational practices characterizing the history of the mint: is it a satisfactory answer to the enigmas centering around the empirical and theoretical dimensions of the disputed transitional significance of the mint and other forms of collective work-place? It must already be clear that the answer should be negative: such dichotomies (one side of which must bear the mark of the industrial revolution to come, the other apparently characterizing an organisational dead-end) shuts out the wide range of different organisational alternatives which the Indian evidence at least serves to illustrate; it involves a far too-rigid categorisation of pure representative, management- (and production-line-) types, conceived as a bifurcation between two opposed evolutionary principles. Needless to say, where early-modern European minting came increasingly under central control, direction and regulation, inhibiting response to changing structures of demand, there is some truth to at least one side of Lane and Müller's formula, while forgery and the manufacture of tokens might supply the other half, although in organisational contexts very different from that of the «machineless factory». The Imperial mint at Agra might also fit the first side of the formula, or that of medieval Venice, where the Commune ostentateously refused to countenance farming of the mint.[110] Indeed at Venice, inflexible working practices and tardy output rates may indeed be connected with the absorbtion into the mint of the individualistic craft practices of the medieval town, but strict regulations limiting supplies of raw materials to the craftsmen (and carefully documented by Lane and Müller), besides imposing a very low cutter/striker ratio (2:1), must seem a much more significant factor in preventing the reorganisation of the workline (while, in turn, it is difficult to comprehend the massive output of 1422–24 without a radical, if temporary reorganisation).

Nevertheless, the same official constraints operated in the farmed mints of the Burgundian Low Countries, official regulations again affecting the proportions of the crafts relative to one another, and ostensibly hedging the freedom of mint-farmers to change the organisation of work beyond the defined limits (although, once more, these limits must have been stretched – even ignored – during recoinages).[111] Burgundy is a particular case in point: many mints in medieval and early-modern Europe represented combinations of official and «private» interventions, a fact even more clearly the case if we include illegal minting, as we should; this was clearly the case in early-modern India, where even some of the most important of the Mughal mints of the 17th and early 18th centuries (Surat, Ahmedabad, Murshidabad, Kasimbazar), demonstrated a strong local involvement by commercial and financial dealers, but which very involvement, in the first case, seemed to

[110] F.C. Lane and R.C. Müller, *Money and Banking*, I, p. 211.
[111] P. Spufford, *Mint Organization*, pp. 239–261.

militate against a unified «conveyor-belt» type organisation.[112] On the other hand, the involvement of private merchants and bankers in establishing, monopolizing or farming mints in 18th century India, was likely to result in the introduction of a variety of different kinds of working practice, depending upon interactions of the various interests concerned with minting, and upon specific local conditions. Conversely, it is equally clear that many official and institutional managements, in a period of rising overall governmental costs, were as concerned to increase mint-output and thus mint-income, as were farmers and money dealers. The point, however, is not to confound attempts to generalize but to seek a kind of understanding appropriate to current sensibilities concerning the complexity of the phenomena of transition, and of the varied evidence for the case in hand.

In this respect, too much attention has traditionally been concentrated upon the precious metals, partly because information about copper and baser billon is scarce, but also because apparently less significant. In fact, because of the role of base monies in extending the social base of money-use and in vastly increasing the scale and depth of monetary economy (and thus of signifiying fundamental shifts in economy itself), they are of very considerable importance, and this importance extends into the mint through generating different working practices, varying combinations of different levels of skill, and widely contrasting output rates, ... all of which - from gimcrackery to quality crafting - are vital to a full perspective of the mint as a collective manufactory, thus of the degree to which the mint was, or was not a forerunner and nucleus of characteristic transitional processes.[113] This implies serious qualification to some traditional ideas on mint trends over the long period, for example the repeated opinion that the period saw a transition towards fine-arts die-engraving from cruder earlier practices.[114] While true for special productions, such as medals and for certain prestige precious metals coinages, the claim involves far too narrow a framework of relevance for identifying the actual range of changes affecting the coin-making crafts, and is thus symptomatic of the whole problem before us. The manufacture in Europe also included the production of dies for an extensive industry in forgery, and in both regions of often very simply marked, official and

[112] However, another influence which needs consideration consists of the technical/architectural restraints on *short-term* expansion in the European mint, namely the grouping of cutting/sizing teams around separate furnaces built into the fabric of the mint (see n. 52, above).

[113] Thus F.C. Lane and R.C. Müller, *Money and Banking*, I, pp. 235-236, briefly distinguish between the working space and conditions accorded, in the new 16th century Venetian mint, to the minting of different metals, copper being confined to a single vault in contrast to the separate shops in which silver coin was factored and among which the division of labour was distributed.

[114] E.g. F.C. Lane and R.C. Müller, *Money and Banking*, I, pp. 239-240.

unofficial base coin. When compared with the relatively smaller output runs of gold coin (thus contrast the shift from gold to silver during the years of the 1640–43 recoinage at Paris), the vigour and volume of copper and billon production seems of particular relevance, because imposing the most severe and untraditional pressures on conventional working practices (and thereby calling into form a variety of institutional responses). In India too, we find this same range of different responses, but almost all departing from older fine-arts standards. By the 18th century, craft quality and gimcrackery form the two essential poles of the universe of workshop and domestic labour, but it is this universe which in both regions is both obscure and obscured.

In this sense, then, it is useful to depart from one of the most influential conventions mutilating analysis of commercial and technical change (and much associated with Weber): the «rationalisation» thesis, together with the supposed evolution of appropriate mentalities, with its curious ethnocentric assumptions and tautologies. In contrast, the range and characteristics of different organisational structures coexisting at a particular time, or distinguishing different periods and places, can best be described in terms of their immediate socio-economic contexts and the institutional constraints locally limiting their response to external stimuli. In these terms, both hand- and machine-production; both disaggregate organisation such as found at Surat, and «conveyor-belt»-type arrangements, such as described at Chandore; both high quality and gimcrack techniques... were each «appropriate» to different combinations of internal and external factors in the same territories during the age of the «commercial revolution». In fact, when we look to the late-medieval and early-modern coin-making crafts, before the introduction of the fully mechanized factory-mint - relating them to the highly pluralistic character of the currency systems of the time - we may then find it worthwhile to describe them in terms of a force-field of consciously mediated alternatives, varying according to both general and local conditions, and responses: it is this force-field which must then be of primary interest, but which is obscured by prevailing methodological habits and general assumptions.

REFERENCES:

A. Fazl Allami. *The A'in-i Akbari*. I. New Delhi, 1977 (reprint of tr. by H. Blochmann. 2nd edn. of 1927 revised and edited by D.C. Phillott).

S. Bhattacharya. *The East India Company and the Economy of Bengal from 1704 to 1740*. London, 1954.

J.J. Brennig. *Silver in Seventeenth Century Surat. Monetary Circulation and the Price Revolution in Mughal India*. In J.F. Richards (ed.). Precious Metals in the Later Medieval and Early Modern Worlds. Durham N.C., 1983 pp. 477–496.

K.N. Chaudhuri. *The Trading World of Asia and the English East India Company 1660–1760*. Cambridge, 1978.

J. Craig. *The Mint. A History of the London Mint from A.D. 287 to 1948*. Cambridge, 1953.

A. Das Gupta. *Indian Merchants and the Decline of Surat c. 1700–1750*. Wiesbaden, 1979.

- - -. *The Merchant and His Money. Glimpses of the Western Indian Ocean in the Early 18th Century*. Paper presented to the Conference on the Monetary System of Mughal India. Durham N.C., 1981.

M. Elvin. *The Pattern of the Chinese Past. A Social and Economic Interpretation*. Stanford, 1973.

Ch. Fawcett. *The English Factories in India 1670–1684*. 4 vols. Oxford, 1936–1955.

W. Foster. *The English Factories in India (1618–1669). A Calendar of Documents in the India Office. British Museum and Public Record Office*. 13 vols. Oxford, 1906–1927.

H. Furber. *Rival Empires of Trade in the Orient 1600–1800*. Minneapolis, 1976.

CH.C. GILLESPIE (ed.). *A Diderot Encyclopedia of Trades and Industry. Manufacturing and the Technical Arts in Plates Selected from L'Encyclopédie, ou Dictionnaire Raisonné des Sciences, des Arts et des Métiers, of Denis Diderot*. 2 vols. New York, 1959.

Gazetteer of the Bombay Presidency. XVI: *Nasik*. Ed. J. CAMPBELL. Bombay, 1883.

K. GLAMANN. *The Dutch East India Company's Trade in Japanese Copper (1645–1736)*. In: Scandinavian Economic History Review 1 (1953), pp. 41–79.

PH. GRIERSON. *Note on Stamping of Coins and Other Objects*. In Ch. SINGER, E.J. HOLMYARD, A.R. HALL and T.I. WILLIAMS (eds.). A History of Technology. II: The Mediterranean Civilizations and the Middle Ages c.700 B.C. to c.A.D. 1500. Oxford, 1956, pp. 485–492.

– – –. *Numismatics*. London, 1975.

J. HABERMAS. *History and Evolution*. In: Telos 34 (1979), pp. 5–44.

I. HABIB. *The Currency System of the Mughal Empire (1556–1707)*. In: Medieval India Quaterly 4 (1961) pp. 1–21.

– – –. *Potentialities of Capitalistic Development in the Economy of Mughal India*. In: Journal of Economic History 29 (1969) pp. 32–78.

– – –. *Monetary System and Prices*. In T. RAYCHAUDHURI and I. HABIB (eds.). The Cambridge Economic History of India. I: c.1200–c.1750. Cambridge, 1982, pp. 360–381.

– – –. *A System of Trimetallism During the Price Revolution. A Study of the Extent of the Use of Silver-Money in the Mughal Indian Economy*. In J.F. RICHARDS (ed.). Imperial Monetary Systems in Early Modern India. Delhi, 1986.

W.W. HUNTER. *Statistical Account of Bengal*. IX: *Districts of Murshidabad and Pabna*. London, 1876.

P.M. JOSHI. *Coins Current in the Kingdom of Golconda*. In: Journal of the Numismatic Society of India 5 (1943), pp. 85–95.

B.A.J. KEMPERS. *Journaal van Dircq van Adrichem's Hofreis naar den Groot-Mogol Aurangzeb (1662).* (De Linschoten Vereniging 45). The Hague, 1941.

F.C. LANE and R.C. MÜLLER. *Money and Banking in Medieval and Renaissance Venice.* I: *Coins and Moneys of Account.* Baltimore, 1985.

J. MALCOLM. *A Memoir of Central India Including Malwa and Adjoining Provinces....* 2 vols. London, 1823.

J. MARSHALL. *John Marshall in India. Notes and Observations in Bengal (1668-1672).* Edited by S.A. KHAN London, 1927.

TH. MARSHALL. *Statistical Reports on the Pergunnahs of...South Maratha Country.* Bombay, 1822.

M. MATE. *Coin Dies under Edward I and II.* In: Numismatic Chronicle (1969), pp. 207-218.

F. MENDELS. *Proto-Industrialization. Theory and Reality. General Report.* Paper contributed to Section A2: Eighth International Congress of Economic History. Budapest, 1982, pp. 69-107.

K.M. MOHSIN. *A Bengal District in Transition: Murshidabad (1765-1793).* Dacca, 1973.

J.H. MUNRO. *Bullion Flows and Monetary Contraction in Late-Medieval England and the Low Countries.* In J.F. RICHARDS (ed.). Precious Metals in the Later Medieval and Early Modern Worlds. Durham N.C., 1983, pp. 97-158.

R.B. OUTHWAITE. *Inflation in Tudor and Early Stuart England.* London, 1969.

J. OVINGTON. *A Voyage to Surat in the Year 1689.* Ed. H.G. RAWLINSON. London, 1929.

F. PERLIN. *Changes in Production and Circulation of Money in 17th and 18th Century India. An Essay on Monetization before Colonial Occupation.* In E. VAN CAUWENBERGHE and F. IRSIGLER (eds.). Minting, Monetary Circulation and Exchange Rates. Akten des 8th International Economic History Congress, Section C7. Budapest 1982. (Trierer Historische Forschungen 7). Trier, 1984, pp. 323-360.

– – –. *Money-Use in Late Pre-Colonial India and the International Trade in Currency Media*. In J.F. RICHARDS (ed.). Imperial Monetary Systems in Early Modern India. Delhi, 1986.

– – –. *Monetary Revolution and Societal Change in the Late Medieval and Early Modern Worlds*. In: Journal of Asian Studies 46 (1986) (forthcoming).

O. PRAKASH. *Foreign Merchants and Indian Mints in the 17th and Early 18th Centuries*. In J.F. RICHARDS (ed.). Imperial Monetary Systems in Early Modern India. Delhi, 1986.

F. PRIDMORE. *The Coins of the British Commonwealth of Nations to the End of the Reign of George VI 1952*. IV–1: *East India Company Presidency Series c. (1642–1835)*. London, 1975.

B.S. PULLAN. *Between Silver and Gold*. In: Times Literary Supplement 4320 (17 January 1986), p. 67.

T. RAYCHAUDHURI. *Non-Agricultural Production: Mughal India*. In T. RAYCHAUDHURI and I. HABIB (eds.). The Cambridge Economic History of India. I: c.1200–c.1750. Cambridge, 1982, pp. 261–307.

J.F. RICHARDS. *Introduction*. In J.F. RICHARDS (ed.). Precious Metals in the Later Medieval and Early Modern Worlds. Durham N.C., 1983 pp. 3–26.

M. DE ROOS. *Sij Ontboode te Coomen Rekenen: Munt en Muntpersoneel in Holland in de Vijftiende Eeuw*. (Projekt Bureaucratisering. Instituut voor Geschiedenis). Groningen, 1978.

H.W. VAN SANTEN. *De Verenigde Oost-Indische Compagnie in Gujarat en Hindustan (1620–1660)*. Leiden, 1982.

D.G. SELLWOOD. *Medieval Monetary Techniques*. In: British Numismatic Journal 31 (1962), pp. 57–65.

F.C. SPOONER. *L'Economie Mondiale et les Frappes Monétaires en France (1493–1680)*. Paris, 1956.

– – –. *The International Economy and Monetary Movements in France (1493–1725)*. Cambridge Mass., 1972.

– – –. *On the Road to Industrial Precision: the Case of Coinage in the Netherlands (1672–1791).* In: Economisch- en Sociaal-Historisch Jaarboek 43 (1980), pp. 1–18.

P. SPUFFORD. *Mint Organization in the Burgundian Netherlands in the Fifteenth Century*. In: C.N.L. BROOKE, B.H.I. STEWART, J.G. POLLARD and T.R. VOLK (eds.). Studies in Numismatic Method Presented to Philip Grierson. Cambridge, 1983 pp. 239–261.

F.W. STAPEL. *Peter van Dam: Beschrijvinge van de Oostindische Compagnie*. I-2. (Rijks Geschiedkundige Publicatiën 68). 's-Gravenhage, 1929.

– – –. *Peter van Dam: Beschrijvinge van de Oostindische Compagnie*. II–3. (Rijks Geschiedkundige Publicatiën 83). 's Gravenhage, 1939.

J. SPLINTER STAVORINUS. *Voyages to the East-Indies by the Late John Splinter Stavorinus*. Tr. and annotated by S. HULL WILCOCKE. London, 1789.

B.E. SUPPLE. *Currency and Commerce in the Early Seventeenth Century*. In: Economic History Review 2nd Series 10 (1957), pp. 239–255.

R. CARNAC TEMPLE. *The Diaries of Streynsham Master 1675–1680 and Other Contemporary Papers Relating Thereto*. I: *The Diary (1675–1677)*. London, 1911.

E. TERRY. *A Relation of a Voyage to the Easterne India, Observed by Edward Terry, Master of Arts and Student of Christ-Church in Oxford*. In S. PURCHAS. Hakluytus Posthumus or Purchas his Pilgrimes Contayning a History of the World in Sea Voyages and Land Travells by Englishmen and Others. IX. [s.l.] 1905, pp. 1–54.

C.R. WILSON. *The Early Annals of the English in Bengal. Being the Bengal Public Consultations for the First Half of the Eighteenth Century*. II–1: *1711–1717*. Calcutta, 1900.

– – –. *The Early Annals of the English in Bengal. Being the Bengal Public Consultations for the First Half of the Eighteenth Century*. III: *1718–1722*. Calcutta, 1917.

E. ZAY. *Histoire Monétaire des Colonies Françaises d'après les Documents Officiels*. Paris, 1892.

MINING ORDENANZAS AND SILVER PRODUCTION AT POTOSI: THE TOLEDO REFORMS

by Alan K. CRAIG

> «... se tiene espiriencia que los hombres rricos no se aplican a labrar Minas por que an bisto que en ello an perdido muchos sus Haziendas y los que tienen poco caudal no se daña buscarlas.»
>
> *Ordenanças de los Comysarios* [No. 6] (Unpublished Archivo Municipal de Arequipa)

Don Francisco de Toledo has been intensively studied by scholars and is consequently recognized as one of the most important early colonial viceroys of Peru and certainly the most efficient administrator. His biographer (Levillier, 1935, p. 13) summarizes his contributions most succinctly:

> «Pero su titulo principal a la gratitud americana, es haber dado normas, fijado principios, protegido al indigena y transformado un virreynato indisciplinado en un Estado de sólida y avanzada estructura.»

Wilgus and d'Eça (1963, p. 66) claim he was the greatest Peruvian viceroy of the sixteenth century buy their opinion was not shared by some of his contemporaries; Garcilaso de la Vega, for example, did not like him at all and others have blamed him for institutionalizing the poorly understood *mita* labor system for mining districts.

After many years of devoted service to the Crown, the aristocratic Toledo was named Viceroy of Peru in 1569 at age 54. He remained there for 12 years, finally returning to Spain in 1581, completely broken in spirit and health, and died shortly thereafter.

During his tenure in Peru, Toledo was instrumental in founding of Huancavelica, Oropesa, Charcas, Cochabamba, Santa Cruz de la Sierra, Tarija, Salta, and Jujuy, all of which survive to this day, but he is best remembered for the promulgation of mining ordinances and his stimulation of this industry in general. As we shall see, not all this credit should be given to Toledo alone as he benefitted from the work of several able predecessors who have not received the attention they deserve.

In the decade after the conquest of Peru only one important silver mining settlement (*asiento minero*) developed in the Andes. At Porco native silver deposits cropped out at the surface and were exploited prehistorically by the

Inca, but were not found by Spaniards until 1540. Spanish mine owners at Porco were the very first to stampede to nearby Potosí in 1545 when one of their Indian workers made the spectacular discovery. Soon thousands of prospectors and miners were swarming over the famous Cerro Rico and surrounding districts in a mining boom without parallel in the New World (Cañete y Domínguez, 1952). As more silver mines were discovered in remote corners of Alto Peru where communications were extremely difficult, the situation became hectic and administration of mining claims can only be described as disorderly.

1. Colonial Mining Ordenanzas

The Iberian Peninsula has never been an important mineral province and medieval Spanish peasantry were not skilled in mining techniques. Nevertheless, a corpus of Spanish mining regulations had evolved that were considered sufficient for the relatively simple geological conditions encountered there. These regulations were transferred without modification to Peru under the assumption that they would serve equally well in the New World. Clearly they did not and it was necessary for President La Gasca to improve and expand on the traditional regulations in an attempt to restore order to the near anarchy at Porco and Potosí.

The ecclesiastic La Gasca was an obedient, temporary administrator for Felipe II and not by any means an innovative or imaginative executive. Toledo was generally contemptuous of him and perhaps a little jealous of his higher religious achievements. However he felt compelled to accept a collection of the old Spanish ordenanzas (Appendix A) enforced by La Gasca, a number of previously unpublished ordenanzas of La Gasca himself (Appendix B) and a few locally derived «Ordenanzas de los Comysarios» (Appendix C). To this heterogeneous mixture, Francisco de Toledo was anxious to add his own contributions that he felt were desperately needed.

The problem was simple. Silver production had crashed after a few flush years following the 1545 discovery. Both native silver and incredibly rich silver oxide ores were concentrated at or near the surface along five major veins that creased the face of Cerro Rico at Potosí. Beneficiation of these high grade ores was a comparatively simple matter of roasting them and the yields were astonishing. Some mine owners became sudden millionaires and so did the King of Spain. As the legal owner of all natural resources in the New World colonies, he permitted miners to extract his silver in return for a fee. Felipe II quickly became accustomed to receive an opulent royal quinto in the form of silver ingots theoretically representing 20% of the total silver production of his loyal vassals in Peru. With the surge of this new wealth, he embarked on a series of expensive socio-military adventures in Europe, only to become uncommonly embarrassed as these tax funds slowly dried up. He inquired and found the mineral wealth of Potosí had apparently declined due to a number of factors and he charged Toledo with the responsibility of improving this situation. His reaction was to visit personally all the major mining districts of Alto Peru and view first-hand the multiple causes of this decline in silver production. As a result of his lengthy inspections and conversations with the «most experienced and ancient» Spanish mine owners at Potosí and elsewhere, he proposed to stimulate mining by issuing a new set of consolidated ordenanzas that would be favorable to all mining interests.

Writing from La Plata in February, 1574, Toledo clearly states (Levillier, 1935, p. 145) in his preamble to the new ordenanzas, his motives for undertaking a complete revision of the existing laws:

> «...porque dado caso que las que habia sobre esta materia hechas por el presidente Gasca y por el conde [Conde de Nieva] y comisarios y por otros Gobenadores, desdicen algunos casos y faltan otros muchos, y aun en los determinados con la variedad del tiempo, y mas experiencia que ahora se tiene, y mudanza de haberes de algunos de ellos fuere necesario que se proveyesen de otra manera.»

and in the same document he describes in detail why the existing Spanish ordenanzas were no longer adequate:

> «...muchas de ellas por la diferencias de los casos no se pueden aclarar a los negocios de estas partes, porque los cerros donde se han hallado las minas de importancia son mas alto y encumbrados que los de España, y en las medidas que se hacen en la superficie de la tierra, hay en lo bajo mucha diferencia, y las vetas decaen tambien á la parte del Sol ordinariamente en tanto grado, que á sesenta y ochenta estados salen de sus cuadras[1] y aun se vienen a incorporar mas con otras....»

Continuing in the same vein, Toledo remarks that underground geometry of Peruvian silver deposits tended to be distinct from anything know in Spain such that:

> «...en las principales como en las ramas que de ellas salen ha de haber diferencia, y pleitos y muy mayores de mayor importancia cuando mas profundas y hondas fueren en el cerro de Potosí.»

Lawsuits over muddled claims and their boundaries could only result in retarding silver production until they were settled. The traditional Spanish tendency toward litigiousness was exacerbated by the obvious wealth at

[1] That is, the veins at Potosí dipped so steeply to the east that 80–120m below the surface, they were outside the claim boundary.

Potosí and Toledo was anxious to prevent these legal tangles and felt obliged to overhaul the entire body of regulations. He wrote (op.cit., p. 144) that:

> «...porque los pleitos que se empezaban a mover con las nuevas labores[2], y los que han tenido hasta aqui cuando el cerro solia tener mas aprovechamento, son de grande inconveniente para el beneficio de los dichos metales.»

But we can also see in this same preamble to the new ordenanzas, a general belief that Potosí silver veins would again increase in richness if it were only possible to deepen the workings. At that time some of the mines were already more than 200 estados beneath the surface, but were not yet flooded («hasta ahora no ha hecho agua»). The problem was the difficulty in gaining access to the working face by laboriously climbing up and down the flimsy leather ladders (*estribos*). Instead, it was deemed necessary to drive in communal adits form lower down on the side of the mountain in such a way as to intercept a maximum number of veins. Since the entire surface of Cerro Rico had been claimed (more than 5000 had been filed by 1574), an ordenanza permitting public access through a community socavon was necessary. This anticipated wealth (which never materialized) was yet another source of increasing legal battles over ownership and boundaries of claims. As mercury became more readily available for amalgamation, there was a veritable blizzard of lawsuits over disposition of ores and tailings long since abandoned as uneconomic.

Since these matters were difficult enough to resolve at the surface, it was imperative in Toledo's view there should be clear rules applicable deep underground. At the same time he was concerned that open pit (*tajo abierto*) mining should be encouraged whenever possible:

> «...para que los indios esten sin riesgo en el beneficio de ellas» (op.cit., p. 146).

Here we have yet another motive in revising the ordenanzas. Don Franciso de Toledo was a man very much concerned with moral responsibility and the welfare of both his own and the King's dependents. The matter of discharging the «royal conscience» was taken seriously by Toledo, if not by the King himself, and he was anxious that the native working conditions be safe and at the same time efficient for the extraction of silver as expeditiously as possible. Obviously it was not always possible to simultaneously meet both requirements and we may safely assume that in actual practice, production took precedence over safety.

Toledo's ambivalent position can be seen in his preamble to the 1574 ordenanzas (op.cit., pp. 146–147) wherein he notes Indian miners were accus-

[2] This is a reference to Toledo's own efforts to introduce the relatively efficient patio amalgamation process he has recently introduced from Mexico.

tomed to buy silver ore from Spanish mine owners and smelt it in native *huayras* furnaces. They refined this silver in small amounts as a kind of cottage industry and used it for various purposes, especially to discharge their annual tribute obligations. Toledo repeats several times that his new regulations will directly benefit these small, part-time miners, but it is also clear that most of the ordenanzas were to encourage the most «Expert and Experienced» of the Spanish who collaborated with him in the promulgation of the new consolidated regulations.

The role of Alvaro Ruiz de Nabamuel, secretary to Toledo and his travelling companion on the tedious «Visita General», has gone unrecognized. He must have provided an invaluable bridge between the turbulent times of La Gasca when Alvaro first arrived in Peru and the more tranquil era of Toledo. Alvaro was in fact already the executive secretary to the Licenciado Castro, Toledo's immediate predecessor (Cook et al., 1975).

Alvaro's handwriting, scribal contractions, and idiosyncracies of spelling are all exceptionally clear and easy to understand and transcribe as they exist in the Arequipa book of mining ordenanzas. From internal evidence established by the page in which his full sign and rubric are penned, we see that Alvaro personally recopied all the ordenanzas of Spain, La Gasca, The Comysarios, and the 1571 set of Toledo (Appendices A, B, C, D) but the extensive mining ordinances promulgated by Toledo in 1574 are in another and more cursive hand. It is not clear where Alvaro's original of the 1574 ordenanzas was deposited - presumably in Potosí, but possibly in Chuquisaca (Sucre), or Lima, or Sevilla. A careful comparison of Levillier's published text of the 1574 ordenanzas which he found in Paris with that bound in with earlier ordenanzas in the Arequipa book shows minor textural differences, substantial differences in the marginalia, and complete omissions of introductory pages evidently missing in the example Levillier transcribed. This of course proves he did not have access to the original and we must presume the Arequipa version is the most complete and authentic, having been dictated and approved by Toledo and written in Alvaro's own hand.

However, we should remember Alvaro himself merely copied those mining ordenanzas Toledo deemed worthy of saving from the Spanish Valladolid Ordenanzas of 1559, and those of President La Gasca and The Comysarios. In total this only amounted to 4, 11, and 3 respectively. From the numbering system of each earlier group (carefully retained by Alvaro) we can be certain that prior to the effort by La Gasca, there were at least 69 Ordenanzas de España theoretically in force in Peru, since that is the highest number we find; at least another 54 were issued by La Gasca himself. By the same reasoning we know The Comysarios must have added at least 11 more. In other words, Toledo may have had as many as 134 pre-existing mining ordinances from which to pick and choose. In 1571 Toledo merely added one ordenanza of his own (Appendix D) which in effect collapsed the previous

134 + (?) down to 18 which he believed were useful and applicable to conditions at Potosí. Finally, in 1574 he issued 73 «new» ordenanzas in 9 Titulos.

From the above we now see that Toledo had an ample base from which to revise the mining laws of Peru and that he had significant help from various sources. Viewed from the standpoint of mining law it would be more appropriate to call him «The Great Condenser» instead of administrator. Where once there were 134, now only 91 remained - 18 of the «old» and 73 of the «new».

2. The Toledo Reforms

As a result of many interviews and the personal knowledge he gained between 1571 and 1574, Toledo made vast improvements in the existing code. From an examination of the short titles of the 1574 ordenanzas (Appendix E) we note that these new laws were grouped according to various subjects which they addresed. At present we have no certain way of knowing how many of these new ordenanzas represent modifications of preexisting versions in force under La Gasca. But the emphasis in *Titulo* 1 on native rights and participation found in the wording of 1a, 3, 10, 16 and the remarkable provision in 7 allowing «foreigners» the privilege of filing mining claims seems strongly Toledean.

The hand of Toledo is also seen in the improvement of ordenanzas allowing more time to file a claim (T1-4), facilitating corrections in claims (T1-8), requiring that significant shaft works be accomplished (T1-11,12), and above all, encouraging the search for mercury (T1-19, 20). The remaining ordenanzas in Titulo 1, and all of Titulos 2, 3, 4, and 7 were designed to clarify the legalistic aspects of filing claims, accurately staking claims, discovery claims and subsequent filings, and the ubiquitous problem of claim jumping under the guise of supposed abandonment. It was certainly Toledo who introduced the ordenanzas in Titulo 9 that both expedite mining lawsuits, and establish heavy fines for frivolous cases.

Titulos 6 and 8 reflect problems that arose due to the extreme number of mining claims filed on Cerro Rico at Potosí. The geometry of silver veins in this highly mineralized mountain was such that the extraction of ore could only continue if communal access to the deepseated deposits could be accomplished by means of an adit or *socabon*. Provisions for such a horizontal tunnel were complicated but the benefits, including dewatering of flooded shafts above, were of great practical value in stimulating increased silver production. These advantages were recognized years before the arrival of Toledo, and indeed several socabones had been driven into the mountain, but problems arose when greedy owners of the surface claim on which the entrance was opened began to charge exhorbitant fees for use of the socabon. The Toledo ordenanzas swept away this practice and provided for construction funding by all who would benefit from use of the adit.

The practical side of Toledo's reasoning is perhaps best expressed in the ordinance recognizing very high altitude mines could not be continuously worked and therefore should be protected from claim jumpers (T7-7) and another environmentally oriented decree giving similar protection to mines not worked during the rainy season (T7-11). We see another practical concern of Viceroy Toledo in the matter of underground safety - he correctly identifies the security of long rawhide leather ladders as the most critical element in raising ore to the surface from shafts that were already hundreds of meters deep. He mandated resting places at convenient intervals for the Indian workers (*apires*) laboriously carrying bags of ore on their backs as they struggled up these dangerous ladders (T5-3).

Regardless of the many reforms the viceroy made dealing with improvements in working conditions, claim ownership, and legal matters, the most significant impetus he gave to silver mining at Potosí involved control of mercury production at Huancavelica and other scattered localities of lesser importance.

From the careful research by Lohmann Villena (1949, pp. 37-57) we know the King of Spain was well aware of the strategic value of mercury deposits discovered near Huancavelica in 1563. Nearly ten years earlier, a successful «cold» amalgamation process had been developed in Pachuca, Mexico by Bartolomé de Medina which greatly stimulated production of silver from Mexican mines. When news of this so-called patio amalgamation method reached Potosí, the Viceroy Cañete wrote to Viceroy Velasco of Mexico requesting two experts be sent to Peru to teach this technique, but for reasons that were never apparent, they failed to achieve the same results. Blame was placed on the poor quality of locally available mercury (but more likely it was in part due to the «bitter» silver ores then mined at Potosí which resisted amalgamation because they were contaminated with antimony, wolfram, and tin minerals).

The Viceroy Cañete, acting on instructions, attempted to control all mercury production in Peru and limited active exploitation to only two mines. He evidently issued mining ordenanzas regulating native working conditions in mercury mines - a concept most writers would ascribe to Toledo. Lohmann also shows us Cañete's regulations were revoked by the Viceroy Conde de Nieva who then proceeded to completely nationalize the mines and required them to be worked by concessionaires. This arrangement proved to be extremely costly and the King charged Toledo with the responsibility of resolving all of the problems surrounding the mines at Huancavelica.

From correspondence (Levillier, 1925) we know Toledo was most reluctant to proceed with total confiscation of mines at Huancavelica other than the «Descubriadora» that belonged to the encomendero Amador de Cabrera, with whom he had protracted struggles over ownership. On advice (possibly from Alvaro Ruiz de Nabamuel), Toledo adopted the position that the mines belonged to the Crown, but only the claim holders would be allowed to work them under strict supervision of appointed officials. He further recommended that prisoners sentenced to be galley slaves should be sent to Huancavelica instead (Lohmann Villena, 1949, p. 42).

At Toledo's request, the Viceroy of Mexico sent Pedro Fernández de Velasco and Jeronimo Piña de Zúñiga to Peru to teach the intricacies of the patio amalgamation process which had previously failed. In August, 1571 Toledo had two large samples of mine tailing from Potosí delivered to Cuzco and in his presence, Velasco revealed his secret methods. For his troubles he was paid a pittance by Toledo who found him otherwise unimpressive. But the success of the experiment so excited the Viceroy that he made his famous statement that he would now arrange the greatest marriage the world had ever seen between the mines of Huancavelica and Potosí!

Now demand for mercury shot up and the response was increased production from the peculiar mercury «placers». Between May, 1571 and February, 1572, at least 2,500 quintales were produced and the price fell from 180 to 40 pesos per quintal. Potosí became a boom town again and economic recovery of the Viceroyalty was complete.

3. Further Evolution of Peruvian Mining Ordenanzas

Pedro Vicente Cañete y Domínguez ([1952], pp. 605-757) provided a remarkable historical analysis of the mining ordenanzas of Peru together with detailed insights into the reasons behind their complex evolution. He is the only colonial chronicler to have done so and as a participant in the events at Potosí, his *Guia de la Provincia de Potosí* (1787) is especially valuable. From his copious notes and references that support the section entitled «Superentendencia de Minas» it is certain he had access to an nearly complete set of the early ordenanzas of Toledo, and at least some of La Gasca, but nowhere does he cite the Comysarios.

On the other hand, Cañete y Domínguez often describes, and makes numerous references to, the mining ordenanzas issued by Toledo's successors (e.g. Viceroys Velasco, Conde de Monclova, Marqués de Montesclaros, Príncipe de Esquilache, Marqués de Guadalcázar, etc.) and makes it clear these men made additions, recompilations, and several revised editions of mining ordenanzas over the years. These later efforts confirm the fundamental value of Toledo's ordenanzas which remained intact throughout this era and even served to inspire modern mining codes (Cornejo, 1958).

Ordenanzas issued after Toledo's time were, for the most part, not designed to stimulate mining activities - rather, they were efforts to proscribe and punish a multitude of ingenious forms of tax evasion and other dishonest activities invented by faustian mine owners of Potosí. After centuries of intense exploitation, most mines were totally exhausted with the result that Indian labor assigned to these mines eventually became the only valuable asset. Thus the mitayos became the targets of many a nefarious scheme and the reputation of the mita rapidly worsened during the remainder of the colonial era.

This gradual decline in mineral wealth was the underlying cause of maltreatment of the mitayos. While there was plentiful silver to be won, Spanish mine owners had nothing to gain by alienating their workers, but this theoretical attitude gradually changed as the economics of ownership became unfavorable. Toledo's ordenanzas institutionalised a rational behavior toward native mine and mill workers, but great administrator that he was, he could not guarantee the silver itself would last forever.

4. Conclusions

As Viceroy, Toledo brought to Peru a determination and tenacity sufficient to accomplish a formidable series of tasks imposed upon him by the Crown. His genius was his ability to reduce complex mining problems by means of practical solutions suited to the peculiar conditions he found at Potosí.

Toledo fomented silver production by controlling the supply of mercury needed for amalgamation, organized a native corvée labor system on a grand scale, and codified mining laws that encouraged prospecting, simplified claims and ownership requirements, and promoted communal work projects among mine owners.

It is in this restructuring of mining ordenanzas that Toledo demonstrated his personal imagination and initiative and for which he deserves much credit. All of the other institutions with which his name is associated were dictated by others or in fact had traditional origins that preceded him.

5. Epilogue

Quite apart from the archival research underlying my discussion of mining ordinances, the participants of this colloquium may find the following data on early coinage of Potosí to be useful, although it carries many caveats.

No definitive monograph in English has appeared on the coinage of Potosí, nor indeed, on any aspect of this world famous mining center. However there are several Spanish texts that treat the Peruvian colonial mints (Lima, Cuzco, Sucre, and Potosí) as part of the rich heritage of Latin American numismatics (e.g. Herrera 1897; Dasi 1950; Burzio 1953; Yriarte 1960; Calbeto 1974). By far the best of these studies is that of Sellschopp (1979) who, as a long-term resident of Peru, was able to devote much of his later life to the collection and analysis of coins from Lima and Potosí. The present study uses as its data base more than 200 numismatic auction catalogs from houses that have consistently offered Potosí coinage during a time span covering the past fifty years. This survey is restricted to non-milled or «cob» coinage from their inception ca. 1575 to 1772 and is based only on specimens that were illustrated by photographs. This latter requirement was necessary because of the notorious incompetence of catalogers to correctly decipher dates, assayers, and mint marks from the often crudely prepared dies and blundered hand-strikings.

The cob coinage of Potosí has never been popular among collectors, presumably because they found the often crudely made coins ugly and esthetically unattractive. In addition, the series has been suspect because of widespread counterfeiting (Craig 1963). These attitudes may be changing due in large part to publicity surrounding the success of treasure salvors along both coasts of Florida. These caches also represent interrruptions in the transfer of precious metals from Latin America to Europe, a theme of this conference.

Cob coins of Potosí cannot be dated with certainty before 1617 when the year of striking was thereafter placed in the legend. Unfortunately, planchets until 1652 seldom were broad enough to include these dates, so we cannot arrive at an index of yearly silver production by analysing catalog offerings for this period. The very few specimens that do bear clear dates are usually specially prepared presentation pieces, not typical of normal strikings, nor indicative of the quantity struck. Obviously the lack of certain dates in catalog sales may not reflect an actual decrease in silver production for the year itself although there are several unexplained intervals that suggest the mint at Potosí was shut down or seriously hampered in some way. A scan of the record indicates few, if any, eight reales were produced during 1659, 1662, 1668, 1671, 1674, 1676, 1678, 1695, 1698, 1700, and especially during the period 1705–1724; another gap occurs form 1728 to 1739. With the exception of 1740, very little seems to have been produced during the 1740's, and with few exceptions this situation prevailed until 1770. The last two years of cob production are fairly well represented but quality of die work reaches a nadir.

How do we explain these observable differences in present day abundance from a colonial mint whose production we would expect to show a gradual deline proportional to the decreasing silver production of Cerro Rico and surrounding mines of the district? Obviously simple attrition should reduce the frequency of older dates tending to offset any real decline in mintage caused by dwindling silver supply. However, we find such long-term trends interrupted by brief intervals such as 1652–1654, 1679–1690, 1724–1727, and 1740 when coins of these dates are comparatively common. The year 1652 marks the beginning of coinage reform and the introduction of new designs. Until very recently, specimens were scarce and seldom offered; beginning in 1983, auction records indicate a cache was placed on the marker and physical condition shows they were from sea salvage. Similar accidents account for the rarity of the 1698–1699 Cuzco coinage, and Mexican mint products carried by the 1715 flotilla. Other examples of shipwrecks that were transporting specie form Potosí include the ill-fated *Atocha Concepcion*, *Maravilla*, and others less well-known that have been looted by modern wreckers. No data are available indicating the amounts involved but they are known to be very substantial.

Throughout the Andes there has been a long succession of coin caches found on land under dramatic circumstances. These «tapados» have become famous fixtures in the lore that surrounds treasure finds and it is certain they have distorted the present survey. For example, discovery of the Cliza hoard near Cochabamba, Bolivia in 1959 acounts for the abundance of fractional cobs of the period 1737–1743 and the eight reales of 1740 which it contained reappeared in this survey as the only illustrated examples.

Similarly, another spectacular discovery occurred in 1979 during demolition of buildings in the colonial sector of Lima as part of an urban renewal project. Thousands of cobs of the early assayer Diego de La Torre (1577–1588) suddenly were placed on the market and radically changed their former status as rarities.

Since no mention of these caches has been published in the formal literature, it may be of some interest to colloquium participants to consider more details of what is undoubtedly the largest hoard of Peruvian mint products ever found on land or sea. The famous «cementerio» tapado was literally uncovered in Lima in 1961 when approximately 60.000 uncirculated eight reales coins spilled out of rotting canvas bags to the delight of gravediggers working in an abandoned sector of the municipal cementery. These specimens do not affect this study as they were all milled coins of the period 1796–1824, but it is interesting to note they were sequestered over a curiously long time span and always selected soon after being struck. These factors suggest a mint employee was involved.

«Creo de utilidad ... estas ordenanzas ... es cosa harto dificil hallarlas para estudio.»
Levillier 1925.

APPENDIX A
Ordenanzas de España

21. Yten ordenamos y mandamos que el primero Hallare o descubriere mina como primer Hallador y descubridor haya primero rregistro y goze de la medida que como tal ha de tener y de estacarse como le pareciere y le estubiere mejor aun que alcance y tome dentro de sus estacas la cata o catas que los demas que despues del binieren obieren hecho con que ante todas cosas Haga estaca fija la qual no pueda dexar ni dexe estacandose o mejorandose como quiera que se estacare o mejorare y los demas que despues del binieren por su orden se an de y estancando y mejoran de Haziendo estaca fija segund d*** y sido ***** binieren juntos breue y sumariamente se auerigue qual fue El primer hallador y descubridor y el que aueriguare ser primero prefiera rreseruando su derecho a salbo al que todauia pretendiere ser primero hallador.

67. Yten por quanto por espiriencia se a uisto que por pleytos y diferencias que se mueuen sobre posesion de minas labor y beneficio de las dichas minas para y se manda cerrar hasta tanto que se auerigue quien tiene mejor derecho y muchas vezes se esta uno y dos años sin labrarse ni beneficiarse lo qual de mas del daño que las partes rreab(r)en es ** notable perjuizio de nuestra rreal Hazienda y de toda la rrepu(bli)ca por tanto para que todo lo susodicho cesse y que las dichas minas no se dexen de labrar y beneficiar tanto tiempo. ordemos y Mandamos que cada y q(uan)do que lo tal acaciere dentro de quarenta dias por el q**** dicho rreyno y no mas. La mina sobre que se litigare este cerrada ante la justicia las partes digan E alegren de su justo y presenten las escrituras y rrecaudos que tubieren y testigos hasta en numero de doze y no mas y con lo que dixeren alegaren y prouaren dentro del dicho termino sin otra mas conclusion y aueriguacion. la dicha justicia lo uea y determine rreserbando su derecho a saluo a la parte contra quien sentenciare la qual labre y beneficie teniendo Cuenta y rrazon por libro con dia, mes, E año del metal que sacare y de las costas y gastos que la labor y beneficio se hizieren dando fianças de mil ducados para que dara Cuenta com pago de lo suso dicho sin bargo de apelacion fuere condenado y se le mandara lo qual se Haga y Cumpla. asi sin bargo de qual quiera apelacion, nulidad, y agrauio que de lo que se determinare y executare se ynterpusiere y si la parte contra quien se sentenciare se tubiere por agrauiado dentro de tercero dia pueda Apelar y dentro de sesenta dias en grado De apelacion, nulidad, E agrauio,

ambas partes sigan su justicia y presenten sus scripturas y rrecaudos y testigos lo que de derecho obiere lugar sigund ***** y con lo que dentro de los terminos sin otra conclusion ni otra prorrogacion dixeren, alegaren, y proua-ren se determine lo que sea justicia y si la sentencia afuere Confirmatoria todabia la parte ***** y a fabor se diere tenga Cuenta y rrazon del dicha Metal que se sacare y de las dichas costas y gastos segun ***** para dar la compago. Si en bia/ ordinaria fuere bencido y condenado a que lado y sin que aya mas grado de suplicacion ny nulidad ni otro Remedio alguno se rremita la causa a la dicha Justo Para que en posesion y propiedad Haga Justo y de la sentencia o sentencias que cerca dello dieren se pueda apliCar conforme a las leyes de nuestro rreynos en grado de apelacion y su ppon (participacion?) seguir su Justicia como les combiniere y si la dicha sentencia fuere rrebocatoria que la tal sentencia se halleuada a pura y deuida Execucion y la parte en cuyo favor se diere sea puesto la tenencia de la dicha mina a la quel(?) la otra parte en cuyo fauor se diere la primera sentencia. De Cuenta com pago de todo lo sacado y procedido de la dicha mina Hasta El dia que se la quitare sacadas las costas y gastos que la labor y beneficio dellos se obieren hecho y este segunddo(sic) tenedor tenga asi mismo Cuenta y Razon por libro Segund les de lo que de la mina se sacare y de las costas y gastos y de los dicho mill ducados De fiancas para que dar a Cuenta com pago si en poseion y propiedad fuere vencido y condenado a quela de y la dicha causa em posesion y em propiedad se rremita a la dicha justicia para que las partes bia ordinaria pueden seguir Su Justicia.

68. Yten ordenamos y Mandamos que qua(n)da y (quan)do que alguno pidiere mina que otro posea quieta y pacificamente y pisiere asi mismo que la dicha mina se cierre que por que El fundamento principal de lo que en tal quaso Se pretende Son los metales que de las dichas minas se sacaran y por que no se dexen de labrar y beneficiar por los daños que dellos se siguen la dicha Justicia. Mande que dentro de treinta dias perentorias(?) citada a la parte de ynformacion Del derecho que tubiere y Que la dicha parte(?)s o ysiere la del contrario, de lo que biere que le combiene y luego pasados los treinta dias pareciendo tener derecho El que pide, Manda al posedor que dend(sic) en adelante tenga Cuenta y rrazon del Metal y plata Que se sacare de la dicha mina y de las costas y gastos que se hizieren segund questa dicho la ordenança antes desta para darla com pago si fuere uenado lo. El Se guarde y cumpla y execute Sin enbargo de qual quier apelacion, nulidad, E agravio que dello se ynterponga y esto hecho proceda por la dicha causa sin dar lugar ni dilaciones de malicia y Haga Justicia.

69. Yten ordenamos y mandamos que cada y quando que se oficieren casos o que se nombraren terceros por las partes so que la dicha justicia los nombrare que los tales terceros ante todas cosas hagan juramento que bien y fielmente dizan y declararen lo que les pareciere y si los dichos terceros no se concertaren en discordia se nombrara otro tercero y si este tal se comformare

con el parecer de alguno de los dichos terceros aquello se guarde y cumpla y execute y si no se conformare y estabieren singulares en todo, o en parte, bayanse nonbrando terceros Hasta tanto que en toda aya dos pareceres conformes y auiendolos se guarde y executen lo que dixeren y sentenciaren

****

APPENDIX B
Ordenanzas Degasca

2. Yten por quitar los pleytos y diferencias que seulen auer entre los dos descubridoreS sobre quien a de gozar El derecho que tiene por que muchas vezes acaece que muchas personas Andan cateando un cerro, ordeno y mando que aquel goze del derecho de descubridor que primero hallo El metal En aquel cerro aunque otra qualquiera sy a encomencado a catear primero. Porque no se puede llamar Mina A quella donde No se ha Hallado metal y si por quaso hallaron El dicho metal Un mesmo dia y no se pudiere aueriguar qual fue El primero que aquel se llame descubridor El que tubiere primero El dicho metal a manifestar ante la justicia auiendo hecho las las [sic] diligencias sobre dichas.

3. Yten que de aqui adelante alguno rregistrare mina [o] beta nueua de plata azogue/ o oro/ o/ otro metal la aya de rregistrar y rregistre dentro de veinte dias como la descubriere ante El alcalde y escribano [?] de las minas guardando en todo lo de mas a cerca de como se ha de hazar El dicho rregistro la forma y orden de la ordenança diez y seis de las nuevas ordenanças de españa cuyo tenor es este que se sigue.

4. Yten que los oficiales rreales se an obligados a tener buenaguarda los tres lados de los rregistros que se les ynbiaren conforme a la ordenança de españa que esta puesta antes desta y que tengan un libro en que asienten por su orden todos los rregistros que obiere provia y que cada año ynbian ante mi o ante El birrey/ o gouernador que por tiempo fuere, Relacion particular de todas las minas que obiere [en] Su provincia y de las que se benefician y el util que se saca de ellas.

5. Yten por que ay en esta prouincia Muchas minas que antes destas ordenanças Estauan rregistrados y no se an labrado y estan ocupaya que se labran no estan rregistradas por la forma que la ordenança tres y *******/ ordeno y mando que dentro de un mes despues de la publicacion de estas ordenanças los duenos de las tales minas ynbiento *** a los oficiales rreales de los rregistros de las dichas Minas conforme a la ordenança tercera destas/ordenanças y la ordenança diez y seis de españa so pena de que se an

perdidas las dichas minas y pasado El dicho termino se pueden tomar y rregistrar por/otra qualquiera persona.

6. Yten ordeno y mando, que si alguno comprare mina dentro de diez dias sea obligado atornalla a rregistrar aunque el bendedor la tubiese rregistrada y en tal rregistro declare de quien la compro y asi mismo ymbie **** del rregistro a los oficiales reales dentro del t(iem)po que se declara ordenança tres para que se sepa de quien se an de cobrar los derechos rreales.

7. Yten que ninguno Pueda uender ni comprar minas sin que primero tenga discubierto metal fijo en ella o que a lo menos este ahondad a tres estados so pena de perder El precio, El comprador y el Vendedor (de) la mina. La qual se aplique para El denunciador y el precio se diuida en tres partes la mitad para la camara y la otra mitad Juez y denunciador.

8. Y Por quanto Mucha Personas no hazen sino descubrir y tomar minas y rregistra llas y se las dexan por labrar de manera que solo sirue El rregistro de estrouar que otros no las labren y su magestad y la rrepublica son defraudados de la provechamiento que dellas podrian tener y aun que esto esta bastantemente proueydo por ordenanças fechas en este rreyno no sean executado/ ordeno y mando que de aqui adelante los que descubrieren minas nuevas las tomaren estacas y Registraren. Las comienen al labrar y continuen y tengan pobladas por la orden y de la manera que esta mandado las ordenanças treinta y cinco y cinCuenta y dos, cinCuenta y tres y cincuenta y quatro de gasca y seis y siete E onze de los comisarios so las penas en ellas con todas su tenor de las quales dichas ordenanças es este que se sigue se sigue.

Ordenanzas Degasca

35. Yten por que se ue por yspiriencia que en todos los asientos de minas se descubren y rregistran muchas minas trayendose metal de las para rregistrar que se halla la haz de la tierra y segund la dispusicion cajas grascrias(?) que hallan parece claro que si las dichas betas que se rregistran se siguiesen se hallaria en ellas metal fijo y por causa que se destacan muchos una beta ninguno quiere dar la cata Su mina para discubrir El dicho metal esperando que esos/otros lo hagan para que despues que biere cierto El prouecho de la dicha beta descubrirle por lo qual todas las demas betas que rregistran se de an de seguir y sea le que muchas dellas siguiendo se sigan tan buenas como las descubiertas y desto a su magestad E a todo El comund biene muy grand daño y podria ser que si aquellos que la rregistraron no pretendiesen El señorio dellas obiese otros que obiese aparejo y boluntad para descubrilas por bien de sobre todo. Ordeno y mando que todas las vezes que alguno biniere con metal a rregistrar minas que todos los que pidieron estacas en la dicha

beta y tomaren mina antes que se Haga El dicho rregistro se obligen todas de dar una cata de diez estados la mina del descubridor, o hasta tanto que se halle metal fijo por que podria ser hallarse antes de los diez estados que cada Uno de su parte siendo obligado El descubridor a pagar la quarta parte de lo que constare a dar la dicha cata y antes de todas cosas El alcalde ante quien se rregistrare la dicha beta tome un fiador que se obligue en forma que si dentro de quarenta dias no fuere dada la dicha cata pagaran todos derechos (suyos?) para la camara de su magestad los quales desde aorado y por eondencidos(?) y desta manera despues de alumbrada la dicha Veta o la dejaren por ynutil no Hallando el dicho metal fixo o biendo que es buena, la siguiran todo Por el gran ynteres e que dello les biene.

52. Yten por que claramente se be El gran daño y perjuizio que biene a su magestad E a todo El comun que las minas no esten pobladas y se labren por acaece muchas Vezes que teniendo una mina la de ay se ba del dicho asiento y de todo El rreyno y la dicha mina no se labra y so color de terminos que piden para yr por tierra mientras de tienen El dominio y profesion de la dicha mina, y si se pudiese tomar pos(sic) despoblada, abria otros muchos labrasen y sacasen prouecho della y los quintos rreales serian aCrecentados y todo El comun se aprouecharia y considerado que la rrazon por que se concede la dicha facultad de tomar mina es por que las pueblan y labran es muy justo que en dejando las despobladas y labrarlas pierdan y qualquier persona las pueda tomar por despobladas por tanto ordeno y mando que despues la ueta Estubiere rregistrada y dada la cata la mina del descubridor Como como (sic) arriba esta dispuesto y puestos los limites y Mojones como esta ordenado y pasados los quarenta dias que para dar la dicha cata, estan concedidos asi. El descubridor como todos los demas Que se obieren estacado la dicha beta dentro de otros diez dias sean obligados a labrar cada uno su mina so pena que si no la labren dentro del dicho tiempo qualquiera las podra tomar por despobladas haziendoles sole ni dades(?) que abaxo se diran.

53. Yten por que muchas personas tienen Muchas minas y no las labren y pretenden que contener un negro /o yndio en ellas estan pobladas y nadie se las pueda tomar y Con esto sin labrarlas defrauden la ordenança sobre dicha por tanto, ordeno y mando que no se llame El labrar las dichas minas si no andubieren siempre quatro yndios en en cada una /o dos negros y no andando los sobre dichos que justamente qualquiera persona la pueda tomar por despoblada E adquirir El senorio en ella como si El la rregistrara y poblara al principio que se descubrio, Con Que haya las diligencias que abaxo se dira.

54. Yten por que muchas Vezes acaee que alguno descubre alguna beta y la rregistra o estaca la otro desCubrio y es tan pobre que no tiene los quatro yndios/ o dos negros arriba contenidos para poder labrar y poblar la dicha mina para no perdella (sic) y podria ser que el sobre dicho por falta de pusibilidad quisiese labrar por su persona la dicha mina, ordeno y Mando que

labrando En ella por su persona propia Se diga tener poblada la dicha mina que asi tubiere.

****

APPENDIX C
Ordenanzas de los Comysarios

6. otrosi Por que se tiene esPiriencia que los hombres rricos no se aplican a labrar Minas por que han bisto que en ello an perdido muchos sus Haziendas y los que tienen poco caudal no se daña buscarlas por faltarles lo necesario y por ser cosa muy conbeniente que se descubren y labren Minas, ordenamos y mandamos que cuando alguna persona descuriere alguna beta y la rregistrare que todos los que en ella pidieren estacas y tomaren minas contribuyen la quantidad de pesos que fueren menester rrepartiendola en ellos por yguales partes para que se de una cata la mina de tal descubridor de ueinte estados, o mas donde hallare El dicha metal y los juezes les den El fauor necesario para que les den yndios para ello de la comarca donde se descubriere y tomando que sea metal la tal mina de la primera plata que se sacare se buelua y rrestituya a las tales personas que lo asi obieren contribuido y si El descubridor de la tal mina o beta quisiere El solo a su costa dar la dicha cata lo pueda hazer con tanto que dentro de treinta dias desde El dia que se rregistrare e declare su boluntad en esto ante El escriuano de minas y dentro de otros dos meses tenga fecha y dada la dicha cata y pasado El dicho tiempo(?) no todos juntas la den como se contienen esta dicha ordenança.

7. Otrosi por quanto la ue tan minas que se an rregistrado el dicho cerro de mas vetas cinco betas principales que estan al la cuchilla del no se aguardado ni cumplido la orden que se da la ordenança treinta y cinco de las que Hizo El dicho nuestro presidente esperando cada uno quel el otro labre y desCubria el metal, esta suspensa la labor y conbiene Que se descubria y bea El metal si lo ay (en) las dichas betas y se de un pozo de veinte y cinco estados la Mina del descubridor y por que los tales descubridores, o los mas por falta de pusibilidad no los dan / ordenamos y Mandamos que todas las personas que estan rregistradas las dichas betas dentro de tres meses despues de pregonada esta ordenança el dicho asiento, Haga la dicha deligencia Comforme a la dicha ordenança de suso y ponga la dicha cata o pozo veinte y cinco estados y el que asi no cunpliere pierda la mina y se pronuncie por baca pasado El dincho termino y qualquiera se la pueda tomar y rregistrar y se la admita El rregistro della y sea obligado a cumplir la dicha ordenança Con aditamento que si dentro de diez dias despues que hiziere El dicho nuevo rregistro la

persona que antes tenia la tal mina y la tenia rregistrada, quisiere contrubuir con los de mas en dar el dicho pozo questa Caso Haziendo obligacion dello sea admitido y se le buelva la mina y la parte y en el lugar donde antes la tenia rregistrada y el que se la auia tomado la dexe y la pueda tomar en otra parte arriba o abaxo donde estubiere baco (sic) de la misma beta y pasado El dicho termino de los dichos diez dias no sea admytido y por Ebitar gasto y pleytos y dilaciones Sobre el tomar de las tales minas declaramos Que con solo El pedimento de que la quisiere tomar y bista del juez y tomar de Como no estan Hechas las dichas diligencias se les apliquen las dichas minas y betas y esto sea titulo bastante para la posesion y proiedad dellas y Mandamos que el alcalde de mynas tenga especial cuydado y diligencia della las dichas betas y Hazer Que se guarde y Cumpla esta Ordenanca.

11. Otrosi Por que muchos De Los que toman minas por despobladas y les son adjudicadas, las dexan estar como las tenian sin las labrar / ordenamos y Mandamos que las personas a quien se les adjudicar minas por despobladas se obliguen luego a dar fiancas en quantidad de cien pesos o dentro de cinCuenta dias daran un pozo de diez esstados a liende (?) de lo que la dicha mina tubiere labrado y si no lo cumpliere los dichos cien pesos se an aplicados la mitad para El ospital(sic) y la otra mitad para la nuestra Camara y qualquier le pueda tomar la dicha mina con la misma condicion y Pena.

****

APPENDIX D
Unpublished mining ordenanzas of Viceroy Toledo[3]

Don francisco Detoledo Mayordomo de su magestad y su bissorey governador y Capitan general destos rreynos y prouincias del piru E presidente Dela abdiencia y chancilleria rreal que por mandado De su magestad rreside en la ciudad De los rreyes & por quanto su magestad tiene proueydo y mandado, por l** cedulas provisiones e ynstruciones que me tiene dadas que con suma diligencia y cuydado procure y de orden que se labren y beneficien las minas de oro, plata, E azogue, E otros metales que ay en estos rreynos como cosa tan ymportante a su rreal seruicio y beneficios. De su rreal Hazienda, bien y conseruacion de los naturales y espagnoles pobladores y Moradores dellos y para Mayor aprouechamiento, E acrecentamiento suyo y por otras causas y efectos en derecados al servicio de dios nuestro señor y defensa E ampliacion

[3] From internal evidence we conclude this preamble was dictated to Pedro de Navamuel, Toledo's secretary, who accompanied him during his *Visita General*. It apparently was written in Guamanga in June of 1571.

de su sancta fe catolica y para lo cumplir y poner en Ex(ecuci)on la bisita [sic] general que uan Haziendo paraproueer y ordenar lo que mas convenga al asiento y estabilidad Destos de sus rreynos y que se an mantenidos em paz y Justicia proueyendo lo que a sido necesario para tomar (e)spiriencia de la rriqueza De las minas De que he tenido noticia que estan descubiertas para que se labren y beneficien y llegado al **** auia hechas algunas ordenanças para los gobernadores pasados que parecio Por entonces se deuian guardar y Cumplir y su magestad de nuevo tiene fechas. Otras auiendome ymformado de personas graves y de espiriencia y tratando dibersas veces sobre los que mas seria necesario probehe rseE ordenarse asi para lo tocante al bien y conseruacion de los naturales que fuesen a la labor y beneficio de las dichas minas como para la buena guarda y cobranca delos quintos rreales De su magestad y acrecentamiento de su rreal Hazienda y para quitar las dubdas, pleytos y diferencias que se rrecrecen y podrian rrecrecer cerca dellos E mirado estos puntos con Mucha Consideracion y teniendo como tengo la cosa presente, Me Parecio que De mas de las dichas ordenanças Que estan hechas se devian de hazer y guardar las ordenancas Del tenor siguiente

Primeramente. ordeno y Mando que en todas las minas deste distrito de guamanga E asientos dellas de plata, azogue, y de otra qualquiera metal asi en las que aora estan descubiertas como en las que adelante si descubrieren se guarden las ordenanças fechas en estos rreynos. Para las minas y los alcaldes de las dichas minas tengan mucho cuydado de las cumplir y executar y si se ofreciere caso que no este dezidido por las dichas ordenanças, se ocurra a la que se Hizieron los rreynos de españa aora ultimamente pa la labor y beneficio de las minas de oro y plata.

Yten que las dycisiones de los pleytos que se ofrecieren sobre posesion /o propiedad de minas o sobre tomarlas por despobladas se guarden las ordenanças veinte y una, sesenta y siete, sesenta y ocho, sesenta y nueue de las ordenanças nueuas de españa y la ordenança quinta de las ordenanças que Hizo gasca para este rreyno con que en quanto la dicha ordenança sesenta E ocho manda que con cada sentencia que se diere los pleytos de posesion de minas se desposea El condenado y se meta en posesion aquel en suyo fauor se obiere dado la tal sentencia. esto se entienda obligandose y començando El que nueuamente se metiere la posesion a labrar la dicha mina con otra tanta gente, herramientas, E aparelos como la labraua El que antes la poseya y de otra manera no sea despo(b)lado. El poseEdor y Con que en quanto por la dicha ordenança se manda que El que entrare la posesion de fianças de Hasta en quantidad de mill ducados no sean estas fianças limitadas sino de todo El ynteres y juzgado y sentenciado y con que en quanto la dicha ordenança se dize que el poseedor tenga cuenta de las costas que Hiziere labor de las dichas minas y del prouecho que dellas acare esta Cuenta la aya de tener Un tercero y no El mismo poseedor y el tenor de las dichas ordenanças Veinte y una, sesenta y siete, sesenta E ocho, sesenta y nueue de españa Y la ordenança quinta de gasca es este que sique.

****

APPENDIX E[4]
Indice de las Ordenzas [sic] de Mineria del Virrey D. Francisco de Toledo

TITULO 1
De los descubridores des Minas

[4] The following index appears in the Arequipa book of mining ordinanzas as an insert in a different hand at a much later date.

TITULO 2
Denuncios de Minas

TITULO 3
Como se han de medir las Minas

TITULO 4
De las quadras y lo que en allas [sic] se guarde

TITULO 5
Como se han de labrar las Minas

TITULO 6
Sobre entradas a Minas a otras

TITULO 7
Del despueble de Minas

TITULO 8
De los Socabones

TITULO 9

Jueces y juicios sobre Minas

REFERENCES:

P. Cañete y Dominguez. *Guia Histórica, Geográfica, Física, Política, Civil y Legal del Gobierno E Intendencia de la Provincia de Potosí* [1787]. La Paz, 1952, 838 pp.

D.N. Cook, A. Malaga Medina and Th. Bouysse Cassagne. *Tasa de la Visita General de Francisco de Toledo*. Lima, 1975, 341 pp.

A. Cornejo. *Las Ordenanzas del Virrey Toledo Como Fuentes del Codigo de Minería Argentino*. In: Revista del Instituto de Historia del Derecho, 9 (1958), pp. 11–13.

R. LEVILLIER. *Gobernantes del Peru Cartas y Papeles Siglo XVI*. Tomo VIII; *Ordenanzas del Virrey Toledo;* (Coleccion de Publicaciones Historicas de la Biblioteca del Congreso Argentino). Madrid, 1925, 408 pp.

Don Francisco de Toledo Supremo Organizador del Peru, Su Vida, Su Obra (1515-1582). Años de Andanzas Y De Guerras. Madrid, [S.D.], 487 pp.

G. LOHMANN VILLENA. *Las Minas de Huancavelica en los Siglos XVI and XVII*. (Publicaciones de la Escuela de Estudios Hispano-Americanos de Sevilla 50. Ser. 2a., 14) Sevilla, 1949, 466 pp.

E.A. SELSSCHOPP. *Las Acuñaciones de las Cecas de Lima, La Plata y Potosí (1568 - 1651)*. Barcelona, 1971, 159 pp.

A. CURTIS WILGUS and R. D'EÇA. *Latin American History*. New York, 5th ed. 1963, 466 pp.

A MICROECONOMIC QUANTITY THEORY OF MONEY AND THE PRICE REVOLUTION[1]

by Kerry W. DOHERTY and Dennis O. FLYNN

1. Introduction

Despite some differences in approach and emphasis, there are basically two camps in the debate over what caused early modern price inflation. The traditional position is that the price rise was caused by the mining of precious metals. This orthodoxy, itself dating from the sixteenth century, is facing the challenge of a growing number of revisionists today. Twentieth-century advocates of the traditional argument have relied almost exclusively on Fisher's quantity theory of money, while opponents typically offer an inflation-via-populationgrowth alternative. The Hamilton-Fisher quantity theory stands on one side; the population thesis stands on the other.[2]

The accumulation of historical evidence continues to undermine Hamilton's quantity-theory view. Cipolla (1955) initiated a lasting attack by pointing to an international chronology problem: Italy's price level rose most rapidly before the influx of Spanish-American metals and it subsided later when the treasure finally arrived in significant quantities. This chronology problem turned up repeatedly in subsequent studies of other countries.[3] Hamilton was unable to respond to this criticism because his own monetary theory, shared with his critics, does require correspondence between international bullion flows and country-by-country price inflation. Empirical

---

[1] Earlier versions of this paper were presented at UC Berkeley, Stanford University, the University of Utah, and the Economic and Business Historical Society. We thank seminar participants for suggestions and criticisms, especially J. Blicksilver, Ch. Calomiris, P. David, E. Randa, D. Roper, R. Sutch, and G. Wright, without implicating any of them in these results.

[2] OUTWAITE'S (1969) survey does a good job of summarizing the Hamilton-versus-Population-Thesis debate up to that time. RAMSEY (1971) contains reprints of some of the classic articles on the price revolution.

[3] See, for example, BLUM (1956), HAMMARSTRÖM (1957), BRENNER (1961), BRENNER (1962), RAMSEY (1971), and DOUGHTY (1975). FLYNN (1978) claims that the same argument can be applied to Spain, but argues that the chronology issue is irrelevant.

studies in the more recent literature provide additional evidence against Hamilton's interpretation; estimated money-supply growth rates, whether traceable to Spanish-American metals or not, do not coincide with movements in countries' price indices.[4] Rejection of the inflation-via-bullion argument is usually coupled with arguments in favor of inflation-via-population growth.

Aside from the perennial need for better data and econometric testing, there seem to be two research alternatives: (1) accept the contention that the precious metals have been dethroned as the primary cause of sixteenth century price inflation, or (2) concoct a better theory for explaining silver's role in causing the century's price rise. This paper attempts the latter by offering a non-Fisherian model which explains the price revolution via the demand for and supply of silver. Bullion is again placed at the forefront, but in the context of a theory which contrasts sharply with the Fisherian quantity-theory hypothesis. Application of the model to the historical issue at hand - sixteenthcentury silver - comprises the main argument of the paper in Sections 2, 3 and 4. Appendix A contains a central mathematical derivation for the model. Appendix B develops a graphical treatment of demand and supply for any storable commodity. The notion of «time» plays a crucial role in the model; we now turn to divergent treatments of time found in the price revolution literature.

2. A Stocks-Flows Distinction

It has been argued elsewhere that the Fisherian quantity theory (MV = PT), which pervades the price revolution literature, is conspicuously absent from modern theoretical debates. (Flynn, 1984a) Today's money supply curve is analytically separable from its demandside counterpart in standard ISLM analysis. The two are superimposed in the same space (Figure 1), but they are distinct. The horizontal axis in Figure 1 is labeled quantity (M/P) - the amount of «real money balances» - with no time dimension. Both money supply and money demand are considered at an instant in time; the former refers to the total stock of money in the system at that point in time, while the latter refers to the aggregate money inventories that the society would desire to hold at various (hypothetical) interest rates (i). Thus, modern monetary theory is couched in terms of «stocks» in the sense that existing inventories and desired inventories are viewed at a point in time. This is the reason that the horizontal axis is labeled without reference to a time di-

[4] See RILEY and MCCUSKER (1983), GOLDSTONE (1984), and LINDERT (1985).

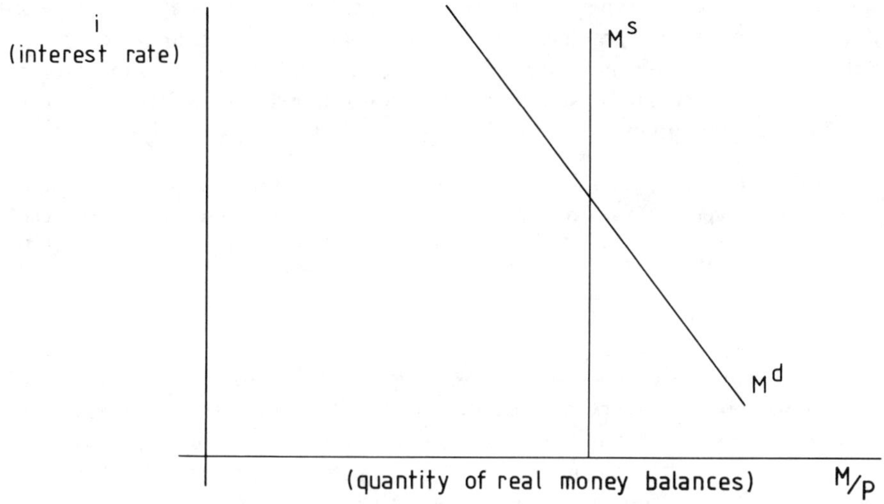

FIGURE 1. Modern ISLM Monetary Theory

mension. Fisher's equation of exchange is formulated in terms of «flows», on the other hand, and it refers to activity occurring over a time interval.

Today's stocks approach to monetary theory differs significantly from the Fisherian-flow rival it defeated early in this century.[5] The fact that there is more than one version of the quantity theory of money is sometimes acknowledged in the historical literature, but differences are normally downplayed as tangential variations within a single tradition. Many consider Fisher's «transactions velocity of money» roughly equivalent to Friedman's «income velocity of money,» for example, but the terminological resemblance masks two fundamental dissimilarities: first, Fisher's velocity concept has to do with money supply, as opposed to Friedman's (Cambridge-k type) money-demand concept of velocity; second, Fisherian analysis utilizes a time interval, as opposed to the point-in-time stocks approach of Friedman. Fisher's transactions velocity of money refers to the number of times a piece of money physically changes hands during a time interval. As interpreted by Blum (1956, p. 198) and others, the money supply is a flow concept (M times V), not just a quantity (M): «Even without any increase in the quantity of money in circulation, the more frequent use of money would have increased

[5] Fisher-type monetary theory failed to explain Gresham's Law early in this century, according to FLYNN and ROPER (1982), who argue that application of stock reasoning eliminates contradictions and generates a theory almost identical with the monetary approach to the balance of payments.

the total of money expenditures. Thus, in effect, the money supply was augmented because of this greater velocity of money.»[6] Friedman's «income velocity,» on the other hand, refers to the portion of wealth held in the form of money (M). His analysis adopts a point-in-time perspective and explicitly emphasizes the demand side:

> «The quantity theory is in the first instance a theory of the demand for money ... It is perhaps worth noting that the model does not use the distinction between 'active balances' and 'idle balances' or the closely allied distinction between 'transactions balances' and 'speculative balances' that is so widely used in the literature.»
> (FRIEDMAN, 1956, pp. 4,14)

All balances are «idle» (i.e. held by someone) at an instant in time, so the active-idle separation serves no useful purpose in point-in-time analysis. The two notions of velocity are distinct - one concerns the money supply over a time interval, while the other refers explicitly to the demand-to-hold money balances at a point in time. Fisher's «flow thinking» should not be confused with the modern monetary paradigm. They reflect contrary traditions in monetary thought.[7]

Hamilton's name is virtually synonymous with this century's bullion interpretation of the price revolution. His reliance on the Fisherian quantity theory has encouraged flow thinking in the price revolution debate; both pro- and anti-bullion camps still rely almost exclusively on Fisher's MV = PT approach.[8] In the case of one important controversy, however, Hamilton proposed an uncharacteristic, non-Fisherian argument in terms of stocks. This atypical side-debate involves a vital theoretical distinction which warrants special consideration.

Hamilton's (1928, p. 34) original article contains a graph which has been reproduced many times since its publication. His data indicated a remarkable correlation between the influx of New World treasure into Spain and the concurrent rise of prices there. (See Figure 2, reproduced from Hamilton,

---

6 Others are less explicit than Blum on this point, but the implication is quite clear that money «not spent» is not, in effect, considered part of the money supply (since MV = O when V = O): «Furthermore, bullion had to be not only imported but also coined, and not only coined but also spent by someone, before having a general impact on prices.» (HAMMARSTRÖM, 1957, p. 124)

7 Acknowledging that Fisher-type monetary theory deserves much of the criticism lodged against it, FRIEDMAN (1956, p. 4) referred to the equation of exchange as «that atrophied and rigid characature.» Friedman's choice of terminology may have unintentionally led others to equate different monetary theories: «Friedman has adopted the Cambridge approach while retaining the velocity terminology.» (ROPER, 1975, p. 15n)

8 Exceptions include MCCLOSKEY (1972) and LINDERT (1985), both of whom utilize stock reasoning.

1934, p. 301.) Hamilton (1928, p. 33) concluded that «[a]mong the causes of the price revolution in Andalusia there can be no doubt that the greatest was the influx of American treasure». In the article's concluding paragraph, however, he addressed the fact that the correlation seemed to break down later in the period, especially after 1630. He explained that non-correlation was to be expected later on because:

> «[e]ven a constant influx of metals would have caused a progressively smaller percentage increase in prices, for each year's imports would have added a smaller increment to the total stock of precious metals»
> (1928, p. 35).

In other words, Hamilton acknowledged that it was the stock of precious metals that mattered, and that the influx of metals was important to the extent that it augmented the existing stock.

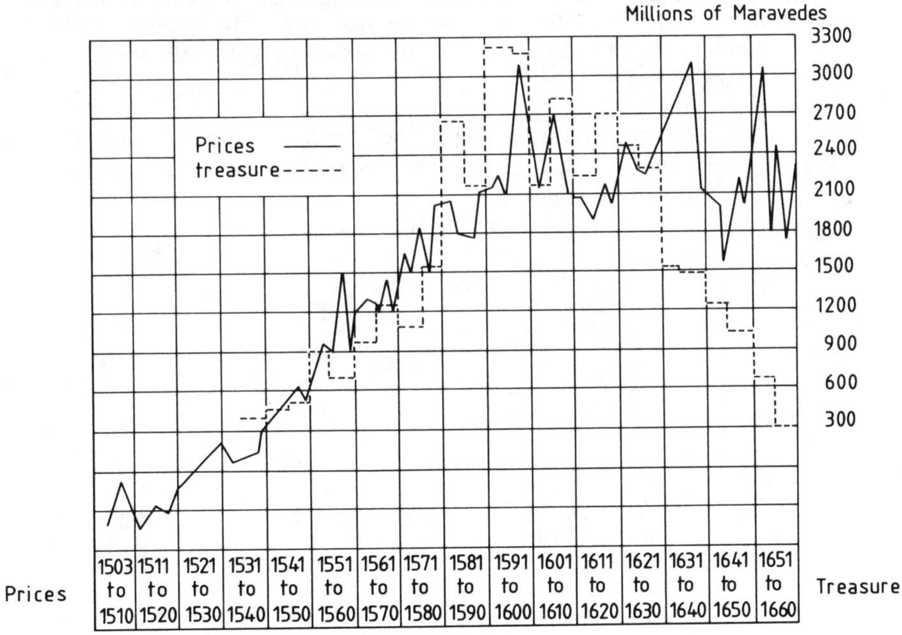

Figure 2. Average Yearly Imports of Treasure to Spain and Index Numbers for General Spanish Prices

SOURCE: Hamilton. American Treasure and the Price Revolution in Spain, 1501-1650, p. 301. Used by permission.

It did not take long for critics to recognize that data on the yearly influx of precious metals need not necessarily correlate with the accumulation of precious metals. How large was the stock to begin with? What about the efflux of metals? Simiand used Soetbeer's estimate as a (1492) base; Simiand concluded that sixteenth-century mine output probably increased the European stock of precious metals tenfold. The problem, then, was that European prices only increased by a factor of three or four (Simiand, 1932 , pp. 436–472). In acknowledging the criticism of Simiand, Hamilton (1934, p. 302) underscored the necessity to discover counteracting forces which would explain «the failure of the [price] upheaval to keep pace with the increase in the stock of precious metals.» Others have likewise recognized that it is the accumulation of metals that is critical, not merely imports from the New World. Vilar (1974, p. 103) posed the issue concisely:

> «Hamilton gives the prices *of Spain* and the arrivals *to Spain*. The revolution of prices is European. If all the metal arrived in Spain and remained in her, the price/metal stock comparison would not be valid for the rest of Europe. But if it does not remain in her, how can one speak of the 'Spanish stock'?... Simiand already observed that what ought to have been represented is not the imports but the growth of the *stock* (that is to say, accumulatively, the existing *stock* at the beginning of each period adding imports and subtracting exports); the latter, however, is poorly known.»[9]

In sum, the accumulated stock of precious metals has been recognized as a central theoretical issue, but estimation of actual magnitudes is dubious at best. Further, the central theoretical issue concernig stocks has never been integrated with the flow (MV = PT) monetary theory which dominates the literature.

---

[9] «Poorly known» is an understatement. There is no possibility of knowing the rate at which specie left most countries in the early modern period. The everyday flow of specie and bullion across borders was not recorded (although some port records exist). There are only guesses about the size of the flow of metals to the Far East, an especially intractable problem when one considers the many overland routes (e.g., via Russia). [Parker (1974, p. 529) and Vilar (1969, pp. 136–137) estimate an annual silver drain of 80,000 kilograms and 64,300 kilograms of silver, respectively, from Europe to the Near and Far East. Neither author reveals a method of estimation. See Chaudhuri (1978) for a classic treatment of the East-West trade; also see Attman (1981) and (1983), who documents the surprising magnitude of the Baltic trade in precious metals.] Even if it were possible to accurately estimate a region's loss (gain) of specie or bullion, one would need to know the «original» size of the stock. But a cursory glance at recent literature surveys of mine production rates (Kellenbenz, 1981) is sufficient to convince everyone, first, of the complexity of the subject (even over shorter time spans) and, second, of the impossibility of estimating world production throughout all time.

3. Modeling Silver at the World Level

The primary purpose of Lindert's (1985) study is to empirically test the Malthusian contention that population growth caused a fall in the English living standard in the early modern period. His findings support the theoretical link between population growth and falling real wages. A central issue in his analysis is the extent to which population growth, compared with monetary growth, accounted for changes in England's price level. Three models provide the basis for his statistical analysis. The first one uses the (stock version of the) quantity theory of money and focuses solely on England. The other two adopt international perspectives. The England-only results indicate that the quantity theory is incapable of explaining a significant portion of England's price revolution. The ensuing «open-economy view» adopts the perspective that (tradables') prices were determined internationally. Even assuming that international monetary forces explain all observed movements in the prices of tradables, the second model again suggests that much of the English inflation cannot be explained by monetary pressure. Finally, a «world-economy view» is discussed. The prices of internationally-traded goods and services themselves require explanation, but hypothesis testing at this level provides a formidable research challenge:

> «This next phase of early modern history will be difficult. Scholars will have to grapple again with the vexing problem of estimating what actually happened to the money supply of western Europe, or the world, as a whole. Existing estimates of the world stock of metallic money are wild guesses for the years before 1800, or even before 1913, as illustrated by the fact that they assume that the stock was zero in 1492.»
> (LINDERT, 1985, p. 629)[10]

The empirical side of a world-economy view does indeed provide many obscure riddles, but replacement of the unsatisfactory Hamilton-Fisher framework is a prerequisite. There is need for a theory which explains the international price level.

Flynn (1983) attempted to model Hamilton's uncharacteristic, non-Fisherian argument on the accumulation of bullion stocks. The plan was to focus on the world market for silver by using basic supply-and-demand theory (see Figure 3).

[10] Some equally wild guesses put Europe's gold and silver stocks at around 3,500 and 37,000 tons, respectively, in the sixteenth century. Using these figures, BRAUDEL and SPOONER (1967, p. 445) estimate growth of Europe's gold and silver stocks at 5% and 50%, respectively, over the period 1500–1650. (This is quite a bit less than the tenfold growth estimate of Simiand reported above.).

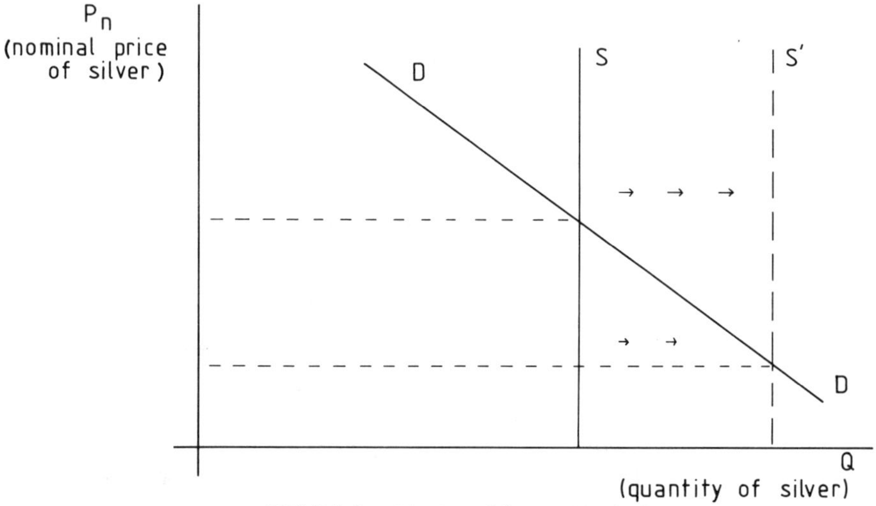

FIGURE 3. World Silver Market

The analysis was intentionally directed at the world level; inter-continental bullion flows testify to the global nature of the market for precious metals. Silver was singled out in the model - rather than, say, gold - because European price data have been systematically reported in terms of «silver-content prices».[11] The vertical axis in Figure 3 refers to the nominal price of silver - the number of monetary units (which depends on which unit of account is used) exchangeable in the market for silver.[12] Choice of nominal price, rather than the ISLM interest-rate label of Figure 1 or the (non-ISLM) reciprocal of the price level as in footnote 12, will be discussed in the next paragraph. «Quantity» on the horizontal axis of Figure 3 does not contain a time dimension; like the (non-Fisherian) model of Figure 1, the analysis describes accumulation (supply) and desired accumulation (demand) at a point in time. Unlike Figure 1, however, the horizontal axis in Figure 3 refers

[11] Wiebe was first to measure changes in «silver-content prices» in 1895. (OUTHWAITE, 1969, p. 24) For a summary of silver-content prices in Europe, see BRAUDEL and SPOONER (1967, pp. 457-486).

[12] Non-Keynesian theorists (e.g. PATINKIN, 1965) have long discussed the market value of money as the reciprocal of the price level; thus, the goods equivalent of a unit of money sometimes labels money's non-ISLM vertical axis. The microeconomic price of money offered here is distinct from the macroeconomic reciprocal of the price level or any vector of the form $(1/P_1,............,1/P_n)$. It is possible that full integration of money and microeconomic theory requires that money's price be dimensioned in the same way as the prices of other goods and services, as is done here - in this case, the price of money is the price of silver.

to the total quantity of silver – monetized and non-monetized combined. Justification for combining diverse forms of silver rests on the assumption of arbitrage among sub-markets.[13] In sum, Figures 1 and 3 both provide point-in-time views of money, but the axes contain different labels. The former is macroeconomic and includes the interest rate on one axis and the quantity of real money balances on the other. Figure 3 is microeconomic with nominal price and quantity of silver (both monetized and non-monetized) labeling its axes.

The nominal price of silver of Figure 3 is expressed in terms of the number of units of some country's money per unit of silver. However, when historical documents have revealed actual, nominal prices expressed in terms of a country's copper, lead, gold, or other coin, it has been customary for price historians to convert these observed prices into silver-content prices. In effect, the silver-content prices record the amount of silver (by weight) that was exchanged for individual commodities. The purpose of this practice has been to permit comparison of various countries' price histories even when actual prices were expressed in terms of differing money substances. Figure 3 does not specify a particular unit of account (in terms of which silver's nominal price is expressed), so it pertains to any country or geographical area. Whatever the (non-silver) unit of account, the nominal price of a good can be converted into a silver-content price.[14] Consider as an example the dimensions of a florin price of wheat (p_j), the jth commodity in a system of n commodities; p_j is the number of florins per hectolitre of wheat. Suppose the nth commodity is silver and its price in terms of florins is denoted p_n, the number of florins per gram of silver. Division of p_j by p_n expresses grams of

---

[13] Silver's value as money could not have diverged, in a persistent and significant way, from silver's value in other uses. When silver's value as bullion exceeded its value as coin by a significant amount, the coins were melted into bullion (thus depleting the stock of specie and augmenting the stock of bullion); bullion was transformed into money, on the other hand, when the mint price exceeded the bullion price. The option of transforming bullion into specie and vice versa guarantees nearly equal prices in sub-markets, thereby permitting conceptualization of a combined silver market. Thus, supply in Figure 3 refers to the world's accumulation of silver (in all forms), while the demand counterpart refers to desired (monetary plus non-monetary) silver holdings. Quantity is therefore not restricted solely to the quantity of money. It is possible that the symmetrical conversion process described here would not apply to silver objects with significant artistic qualities; bullion and mint prices might never rise enough to induce melting and conversion of such objects (just as the price of paint may never rise sufficiently to induce destruction of a Rembrandt painting). We abstract from that portion of the world silver stock that presents this sort of asymmetry and thank Richard Sutch for bringing it to our attention.

[14] The unit of account here cannot be a silver money because it is the market itself that is under study. Thus, silver's bullion price would not have been expected to fall vis à vis a silver money because they share a market value; that is, the purchasing power of silver coins fell as much as silver bullion.

silver per hectolitre of wheat. Thus, p_j/p_n is the silver-content price of wheat. This theoretical conversion describes the silver-content prices obtained in practice by price historians. Each commodity's florin price is a component of the nominal price vector (p_1, p_2, ..., p_j, ..., p_n). Division of each component by p_n generates the relative-price vector (p_1/p_n, p_2/p_n, ..., p_j/p_n, ..., p_{n-1}/p_n, 1); all nominal prices are thereby converted into silver-content prices.[15] This formulation, in conjunction with the demand/supply analysis of Figure 3, will now permit a straightforward explanation of silver-content price inflation in the early modern period.

Following the reasoning of both Classical analysis and Neoclassical microeconomic theory, demand and supply determine the exchange value of a commodity, but price eventually gravitates toward cost of production.[16] New mine discoveries and technological innovation significantly reduced silver's cost of production, creating one of history's great mining booms. The existing stock of silver must have been so great (as Hamilton argued), however, that unprecedented rates of production only slowly increased it. An increase in a commodity's stock supply, other things equal, reduces its market price. Using the analysis developed above, a reduction in the market price of silver (p_n) raises each one of the silver-content prices.[17] We contend that sixteenth-century price inflation resulted when the world's stock supply of silver grew more rapidly than did stock demand. A variety of additional relative-price movements had to have been caused by demand and supply shifts in the other n-1 markets[18], but it was the silver market that accounted for the underlying upward trend of silver-content prices in Western Europe, the Baltic, Russia, the Middle East, and even China.[19] The demand/supply ap-

[15] PATINKIN (1965, Chapter II) used a similar method to discuss the nominal price of the medium of exchange. His model envisions two distinct monies: the (fictional) unit-of-account money is external to the system and serves as numeraire, while the price of the (actual) medium of exchange money is expressed in terms of the unit of account.

[16] For a contrast of the «Classical», «Neoclassical», and «Fisherian» quantity theories, see MASON (1974) and MASON (1977).

[17] It is claimed that silver-content prices rose most rapidly in Spain. According to HAMILTON (1934, p. 202): «To be specific, the average price level was 4.32 times as high in 1601 as in 1501. Since silver prices never fully descended from the peak attained at the close of the sixteenth century, it seems correct to say that by 1600 silver prices in Spain had risen permanently to a level four times as high as that at which they stood in 1501.» A 432 percent rise for the hundred-year period correspond approximately to a 1.5 percent annual rate of price inflation.

[18] As has been argued by ROBINSON (1959, p. 54), HAMILTON (1960, pp. 153–155), and MCCLOSKEY (1972, p. 1333), one cannot use microeconomic demand/supply arguments for the ith commodity (say, wheat) to explain changes in the general price level. It is clear in our model that a change in the price of silver (p_n) has a general impact on all n-1 silver-content prices - that is, it accounts for general silver-content price inflation - while the same is not implied by a change in p_i, $i \neq n$.

[19] On the simultaneity of silver-content price inflation in various regions, see, for example, BARKAN (1973, p. 73), BLUM (1956, p. 185), BRAUDEL (1966, I, pp. 518–522),

paratus of Figure 3 has a number of advantages over previous monetary explanations of sixteenth century price inflation. Among others, it explains the beginning and end of the inflationary episode - lower cost of production due to mine discoveries and technological advance - and the mining surge can be explained via the profit motive.[20]

4. An Inventory-Theoretic Approach to Demand and Supply

The analysis of Figure 3 differs from both Fisherian and modern quantity theories, as well as ISLM analysis, in that the quantity axis refers to monetized and non-monetized silver combined. The isolated object is silver, not money per se. This distinction is noteworthy because focus on a particular commodity - regardless of the number of end uses - is a characteristic of microeconomic analysis. Utility analysis is a cornerstone of microeconomics. Since utility is maximized subject to an income constraint in comparative-static analysis, the dimensions of the horizontal axis in conventional demand and supply - quantity per time period - reflect the length of the income period. Demand and supply are time-dimensioned.

The problem is that the issues discussed in Figure 3 concern inventories at a point in time; its horizontal axis is labeled «quantity» without a time dimension. The stock supply curve refers to the quantity of the item accumulated up to that point in time; it must be vertical because quantity is fixed at an instant irrespective of price. The absence of a time dimension associated with the stock supply curve, however, implies that it cannot be superimposed on the time-dimensioned space of a conventional microeconomic demand curve. Flynn's (1983) model erred precisely on this point. Since stock supply cannot be superimposed with a conventional time-dimensioned demand curve, derivation of a point-in-time demand curve is required. Formal derivation of stock demand - by means of utility maximization subject to a wealth constraint - is presented in Appendix A. Graphical treatment of the interplay

---

BRAUDEL and SPOONER (1967, pp. 397-398), FRIIS (1953, p. 210), GLAMANN (1971, p. 457), HAMILTON (1960, p. 152), HAMMARSTRÖM (1957, p. 120), HOSZOWSKI (1961, pp. 445-446), MALOWIST (1958, p. 31), ZIMYANI (1975, p. 402). Note the intriguing calculations of CARTIER (1981), which show a revolution in the silver-content price of rice in China along the same lines as wheat prices in the West. Some Chinese actually experienced this price inflation, even though their unit of account was linked to bronze, because silver was sometimes used as a medium of exchange.

[20] SZASDI (1981) contains an excellent description of the vagaries and variability of profits in Spanish-American silver mines. FLYNN (1982) and (1984b) argues that these profits financed the Spanish Empire, which declined along with mine profits.

between stock demand/supply and flow demand/supply is contained in Appendix B; the analysis there refers to any storable, consumable commodity. As for early-modern silver, the accumulation portion of Hamilton's thesis is salvageable when viewed from the stock-flow perspective of this microeconomic model.

The market stock supply curve (S) of Figure 4(b) represents the accumulated world stock of silver at some point in time prior to the onset of the price revolution. The actual size of this stock is unknown, and probably unknowable, but suppose it was Q^* in Figure 4(b). Market stock demand (D) represents the sum of the individual inventory-theoretic demand curves derived in Appendix A; it refers to the world market (utility-maximizing) holdings of silver at every price. Equality between stock demand and stock supply determines the market-clearing price of silver at any point in time. Figure 4(a) depicts the dynamic analogue to comparative-static supply; the set of optimum production rates is called «flow supply» (S_{fl}). Silver's rate of consumption is assumed to have been zero; thus, its «flow demand» curve (D_{fl}) adheres to the vertical axis in Figure 4(a).[21]

Consider the following situation. The market-clearing price (p_c) has settled to silver's average cost of production (COP) and has induced a particular rate of production (q_c) exactly equaling the rate at which the stock of silver is worn out and lost (defined as the «decay rate»). These circumstances imply a steady-state world silver stock, other things equal, because the rate of world production of new metal offsets the rate at which the existing stock is depleted. Producers would neither enter nor leave this stable industry because zero economic profits obtain. The world silver market is in equilibrium.

Technological advance created a European silvermining boom commencing in the middle of the fifteenth century, followed by the famous American and Japanese discoveries and further mining innovations during the next century and a half.[22] Ignoring the immense complexity of economic and political interactions surrounding these phenomena, consider only the reduction in the cost of producing silver (to COP'); the new flow supply curve (S_{fl}') represents the collected shift in the rate of world silver production. Would silver's market-clearing price have dropped to its new equilibrium level immediately following the technological advance? The answer is no. The interaction of stock demand and stock supply determines the

---

[21] This is not to deny that silver was lost and worn out. We consider these sources of inventory depletion under «decay» rather than consumption, however, because consumption is restricted to utility-generating activities. Wastage may be necessary, but it is not satisfying. Doherty and Flynn (1986) contains a more complete mathematical treatment of the model described in Appendices A and B.

[22] See especially Nef (1941), Iwao (1959), Kobata (1965), and the essays in Kellenbenz (1981).

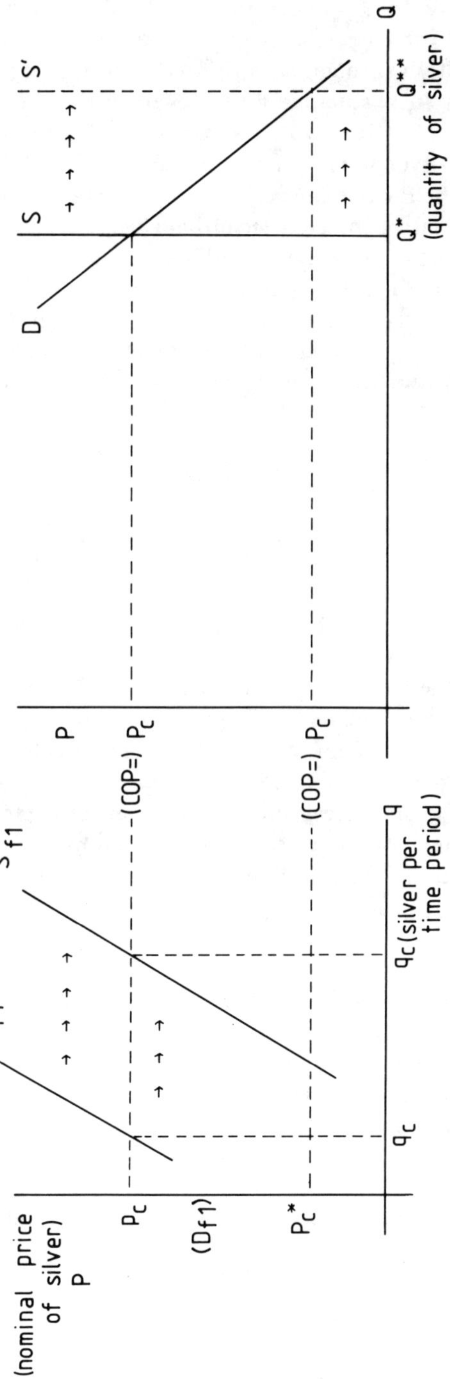

FIGURE 4. A Stocks-Flows Approach to Silver

market price. The difference between the new production rate (say, q_c') and the decay rate would increase stock supply over time, but the initial stock represented accumulation throughout all previous history. Stock supply must have grown gradually in percentage terms, even in the face of an unprecedented rate of production. This would account for the deliberate decline in the market value of silver during the price revolution. The gradual rise of price levels throughout Europe and elsewhere - expressed in silver-content terms - reflects the relative mass of accumulated world silver. When silver's market price finally settled to its new equilibrium level - at its new, reduced cost of production (COP') - the price revolution was over. The purchasing power of silver remained relatively stable for the remainder of the seventeenth century. Further discoveries and technological innovations reestablished disequilibrium during the inflationary eighteenth century, but that is another story.

5. Summary and Conclusions

A growing mass of historical evidence contradicts Hamilton's quantity-theory interpretation of the price revolution. The anti-bullion criticism initiated by Cipolla in the mid-fifties - that the flow of American treasure to individual countries did not match their respective inflationary experiences - received no rebuttal from Hamilton.[23] Moreover, recent empirical work suggests that growth rates in countries' money supplies did not match domestic inflation rates in Europe, irrespective of whether monetary growth was of Spanish-American origin. Weaknesses in the Hamilton-Fisher hypothesis are undeniable. Yet the traditional contention that silver mines explain early-modern price inflation is supported by the inventory-theoretic reformulation of this essay.

In a solitary side-debate, Hamilton adopted an uncharacteristic argument formulated in terms of stocks rather than Fisherian flows. He saw no contradiction between record infusions of bullion into Europe late in the sixteenth century and slower simultaneous rates of price inflation. The annual influx of precious metals increased during the century, but larger absolute additions translate into smaller percentage growth rates because of the increasing size of the stock. A slower growth rate of the bullion stock, according to Hamilton, implied slower growth of the monetary stock. Thus, the falling rate of growth of the bullion stock explains the subsiding rate of price inflation.

[23] Hamilton (1960) reaffirmed all of his major arguments in Stockholm, but he ignored the chronology criticism.

The model of this paper incorporates features of Hamilton's accumulation argument, but there are some significant differences. Rather than focus on money supplies and individual countries, the analysis concerns world demand for and world supply of monetized and non-monetized silver combined. Silver is viewed from a microeconomic perspective. The decline in its market value (i.e. rise of silver-content prices) commenced when stock supply began increasing more rapidly than demand. Mine discoveries and technological innovation explain rising production rates via the profit motive - because production costs had dropped. Generations of high production rates steadily augmented stock supply and thereby depressed the price of silver. The price revolution (in silver-content terms) ended when the world market value of silver stabilized at its new, lower cost of production.

The statistical analyses of Riley and McCusker (1983) and Lindert (1985) suggest that too much emphasis has been placed on the supply-side of the monetary argument; the demand side has been slighted.[24] The stock-flow model of this paper attempts to give the demand side its theoretical due. The interaction of demand with supply explains the episode of silver-content price inflation. Where non-debased silver standards obtained, like sixteenth century Spain, silver-content price inflation was identical with nominal price inflation; the purchasing power of silver coin fell along with that of silver bullion. Deviations from this underlying silver-price trend occurred in areas experiencing debasement, like France, and in regions not on a silver money standard, like China.[25] The purchasing powers of these monies ultimately depended on the cost of producing the substances to which they were linked.

---

[24] When applied to the case of sixteenth-century silver, the stock-flow model of this essay suggests that secular price inflation (in silver-content terms) was ultimately determined by the white metal's cost of production. Empirical estimation of the demand and supply curves themselves may not be possible, but this is true for many commodities. The implausibility of accurate estimation of worldwide silver accumulation has already been discussed. It is clear from the mine production estimates in Kellenbenz (1981) that estimation of flow supply (production) curves are also improbable. Empirical work on estimation of a stock demand curve - the worldwide inventory that people would desire to hold at every price - would be equally difficult. So, useful estimates of silver's worldwide stock supply and stock demand at any point in time are improbable, but evidence of mine discoveries and a series of important technological advances is abundant and incontrovertible. Elusiveness in estimation notwithstanding, demand and supply analysis is an indispensible tool of economic theory.

[25] Thus, Einaudi attributed roughly half of the 627% French price revolution to (a fall in the value of) precious metals and the rest to a halving of the intrinsic content of the French monetary unit. See Braudel (1974, p. 518) and Spooner (1968, p. 19).

APPENDIX A
Derivation of Stock Demand

Let $U = U[Q_1(t), Q_2(t), q(t)]$ be a concave utility function twice differentiable in all arguments with first-order partial derivatives nonnegative, and second-order own partials nonpositive, where

$Q_1(t)$ = stock of commodity 1 (financial asset) at time t
$Q_2(t)$ = stock of commodity 2 (consumable) at time t
$q(t)$ = consumption rate of commodity 2.

Let $Q_1(0) = Q_1^0$ and $Q_2(0) = Q_2^0$ be the initial endowments, and suppose

$\dot{Q}_1(t) = -z(t)$
$\dot{Q}_2(t) = p_1 z(t)/p_2 - q(t)$

where z(t) is the rate at which commodity 1 is used to purchase ($z > 0$) or sell ($z < 0$) commodity 2 at time t, and p_i is the (constant) price of commodity i, i = 1, 2.

The goal is to maximize the present value of discounted utility over an indefinite time horizon by appropriate choice of z(t) and q(t). Thus,

$$\max_{z,q} \int_0^\infty U[Q_1(t), Q_2(t), q(t)]\, e^{-\rho t}\, dt$$

subject to $\int_0^t z(\tau)\, d\tau \le Q_1^0$ and $\int_0^t \left[\frac{p_1 z(\tau)}{p_1} - q(\tau)\right] d\tau \le Q_2^0$

and the other conditions just outlined above.

The Lagrangean function becomes (employing a current-value Hamiltonian):

$$\begin{aligned} L\,[z(t), q(t), Q_1(t), Q_2(t), \lambda_1(t), \lambda_2(t), \xi_1(t), \xi_2(t)] = \\ U[Q_1(t), Q_2(t), q(t)] + \lambda_1(t)[-z(t)] + \lambda_2(t)[p_1 z(t)/p_2 - q(t)] \\ + \xi_1(t) Q_1(t) + \xi_2(t) Q_2(t). \end{aligned}$$

Using Pontryagin's maximun principle, the following denote first-order conditions for maximizing the integral above (time arguments are suppressed for simplicity):

(1) $\dot{Q}_1 = -z$

(2) $\dot{Q}_2 = p_1 z / p_2 - q$

(3) $\frac{\lambda_1}{p_1} = \frac{\lambda_2}{p_2} = \frac{U_q}{p_2}$

(4) $\dot{\lambda}_i = \rho\lambda_i - U_{Q_i} - \xi_i, \; i = 1, 2$

(5) $\xi_i \geq 0, \; Q_i \geq 0$ and $\xi_i Q_i = 0$

(6) $\lim_{t \to \infty} e^{-\rho t} U_q \geq 0$ and $\lim_{t \to \infty} e^{-\rho t} U_q Q_i = 0, \; i = 1, 2.$

Assume that it is optimal for the individual to hold positive inventories for both commodities at time t. In that event, $\xi_1 = \xi_2 = 0$ in (5).
Noting that prices are treated parametrically, take a time derivative in (3) to arrive at

(7) $\dot{\lambda}_1 / \dot{\lambda}_2 = p_1 / p_2 .$

Substitute (4) into (7) to obtain

(8) $\frac{p_1}{p_2} = \frac{\rho\lambda_1 - U_{Q_1}}{\rho\lambda_2 - U_{Q_2}}$

Finally, substitute $_1$ = $(p_1/p_2)\ \lambda_2$ from (3) into the numerator of (8) from which follows

(9) $\frac{U_{Q_1}}{p_1} = \frac{U_{Q_2}}{p_2}$

This is the familiar condition for utility maximization in conventional theory, except note that the conditions are couched in terms of stocks rather than flows, since U_{Qi} is the marginal utility of the stock of commodity i.
Thus, the demand functions generated from (9), given the optimal consumption rate q*, are stock demand functions, written:

(10) $Q_1 * (t) = h(p_1, p_2; Q_1^0, Q_2^0), \; i = 1, 2.$

APPENDIX B
Graphical Treatment of a Storable Consumable

The model in Appendix A describes how an individual simultaneously adjusts consumption rates (q) and inventory holdings (Q_2) in an optimal (utility-maximizing) fashion at each point in time, given a set of prices. For each price p_2, the simultaneous solution to equations (1) through (6) generates optimal q* and Q_2* (among other variables). The locus of (p_2, q*) and

(p_2, Q_2^*) combinations defines, respectively, the individual's consumption (flow) demand curve and inventory (stock) demand curve. The market (flow and stock) demand curves of Figure B.1 are derived in the usual way, by aggregating quantities over the set of individuals in the marketplace. Superimposed in the diagrams are market flow supply (optimal production rates aggregated over the set of firms in the marketplace) and stock supply (the existing stock) for each price p_2.[26] Desired inventory holdings (stock demand) equal existing inventories (stock supply) at the market-clearing price (p_c in Figure B.1).

Abstract from decay and consider the following steady-state scenario. The market-clearing price (that which equates stock supply and demand) also equates flow supply and demand and there are normal rates of return throughout the industry. Thus, there is no excess of consumption over production or vice versa. The existing stock is the desired stock of the commodity and price has dropped to the minimum unit cost of production (assuming competition). Now suppose that circumstances change such that there is a general desire to hold larger inventories at every price (say, because of fear of war) – the stock demand shifts rightward to D'. The market-clearing price (p_c') would then exceed that which equates flow demand and supply. This implies that the market rate of production (flow supply) would exceed the market rate of consumption (flow demand). Items which are produced but not consumed must augment inventories; in other words, stock supply would shift out over time. Stock supply would continue to shift out, putting downward pressure on market price, until price once again equates the market rates of consumption and production.[27] The same logic applies if market price falls below the intersection of flow demand and supply; the excess of consumption over production reduces stock supply, putting upward pressure on market price. A price that equates both stock and flow demand is necessary, assuming zero decay, for a steady state.

[26] The optimal production rates referred to here are discussed in the literature on the theory of the firm. See, for example, TAKAYAMA (1974).

[27] These graphs contain a comparison of two different steady states; rigorous derivation of variables' dynamic adjustment paths is beyond the scope of this paper.

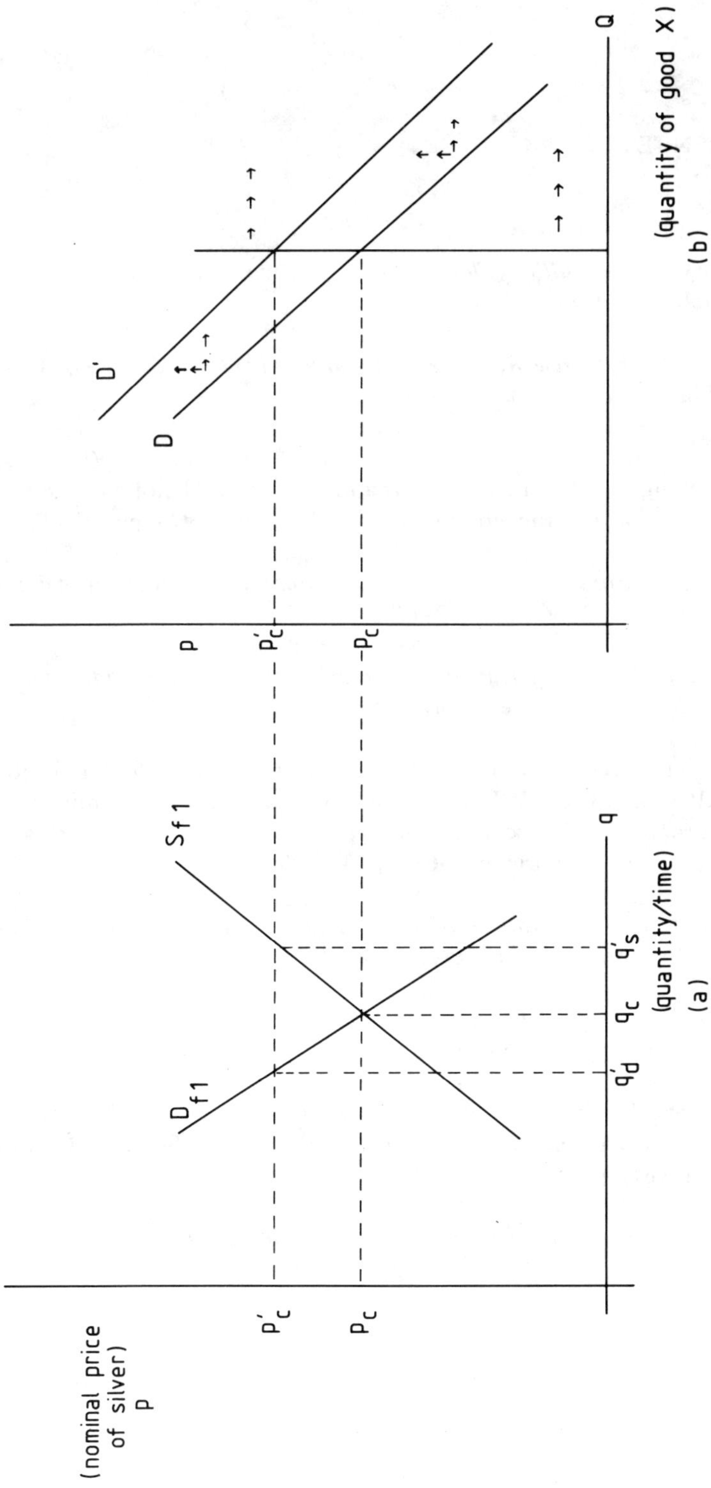

FIGURE B.1. A Storable Consumable

REFERENCES:

A. ATTMAN. *The Bullion Flow between Europe and the East (1000-1750).* Goteborg, 1981.

- - -. *Dutch Enterprise in the World Bullion Trade (1550-1800).* Goteborg, 1983.

O. BARKAN. *Les mouvements des prix en Turquie entre 1490 et 1655.* In: Mélanges en l'honneur de Fernand Braudel. I: Histoire économique du monde méditerranéen (1450-1650). Toulouse, 1973, pp. 65-79.

J. BLUM. *Prices in Russia in the Sixteenth Century.* In: Journal of Economic History 16 (1956), pp. 182-199.

F. BRAUDEL. *The Mediterranean and the Mediterranean World in the Age of Philip II.* 2 vols. New York, 1966.

F. BRAUDEL AND F. SPOONER. *Prices in Europe from 1450 to 1750.* In: E.E. RICH and C.H. WILSON (eds). The Cambridge Economic History of Europe. IV: The Economy of Expanding Europe in the 16th and 17th Centuries. Cambridge, 1967, pp. 374-486.

Y.S. BRENNER. *The Inflation of Prices in Early Sixteenth Century England.* In: Economic History Review 14 (1961), pp. 225-239.

- - -. *The Inflation of Prices in England (1551-1650).* In: Economic History Review 15 (1962), pp. 266-284.

M. CARTIER. *Les importations de métaux monétaires en Chine. Essai sur la conjoncture Chinoise.* In: Annales: Economies, Sociétés, Civilisations 10 (1981), pp. 513-516.

K.N. CHAUDHURI. *The Trading World of Asia and the English East India Company (1660-1760).* Cambridge, 1978.

C.M. CIPOLLA. *La prétendu 'révolution des prix'. Réflexions sur l'expérience italienne*. In: Annales: Economies, Sociétés, Civilisations 10 (1955), pp. 513–516.

K.W. DOHERTY AND D.O. FLYNN. *Consumer Demand in Comparative-Static and Dynamic Analyses*. mimeo (1986).

R.A. DOUGHTY. *Industrial Prices and Inflation in Southern England (1404–1640)*. In: Explorations in Economic History 12 (1975), pp. 177–192.

D.O. FLYNN. *A New Perspective on the Spanish Price Revolution. The Monetary Approach to the Balance of Payments*. In: Explorations in Economic History 15 (1978), pp. 388–406.

– – –. *Fiscal Crisis and the Decline of Spain (Castile)*. In: Journal of Economic History 42 (1982), pp. 139–147.

– – –. *Sixteenth Century Inflation from a Production Point of View*. In: N. SCHMUKLER (ed.). Inflation through the Ages. Economic, Social, Psychological and Historical Aspects. Brooklyn, 1983, pp. 157–169.

– – –. *The Population Thesis Versus Economics and History* and *Use and Misuse of the Quantity Theory of Money in Early Modern Historiography*. In: F. IRSIGLER and E. VAN CAUWENBERGHE (eds.). Münzprägung, Geldumlauf und Wechselkurse/ Mintage, Monetary Circulation and Exchange Rates. (Trierer Historische Forschungen 7). Trier, 1984, pp. 361–382 and pp. 383–419. [1984a]

– – –. *El desarrollo del primer Capitalismo a pesar de los metales preciosos del nueco mundo. Una interpretation anti-Wallerstein de la España Imperial.* In: Revista de Historia Economica II. Primavera-Verrano, 1984, pp. 29–57. [1984b]

D.O. FLYNN AND D.E. ROPER. *Gresham's Law and the Modern Theory of the Demand for Money*. In: Eastern Economic Journal 8 (1982), pp. 219–235.

M. FRIEDMAN. *The Quantity Theory of Money. A Re-statement*. In: M. FRIEDMAN (ed.). Studies in the Quantity Theory of Money. Chicago-London, 1956, pp. 3–21.

A. FRIIS. *An Inquiry into the Relations between Economic and Financial Factors in the Sixteenth and Seventeenth Centuries*. In: Scandinavian Economic History Review 1 (1953), pp. 193-241.

K. GLAMANN. *European Trade (1500-1700)*. In: C.M. CIPOLLA (ed.). The Fontana Economic History of Europe. II: The Sixteenth and Seventeenth Centuries. Glasgow, 1971.

J.A. GOLDSTONE. *Urbanization and Inflation: Lessons from the English Price Revolution of the Sixteenth and Seventeenth Centuries*. In: American Journal of Sociology 89-5 (1984), pp. 1122-1160.

E.J. HAMILTON. *American Treasure and Andalusian Prices, 1503-1660: A Study in the Spanish Price Revolution*. In: Journal of Economic and Business History 1 (1928), pp. 1-35.

- - -. *American Treasure and the Price Revolution in Spain, 1501-1650*. Cambridge, 1934.

- - -. *The History of Prices before 1750*. In: XIe Congrès International des Sciences Historiques. Rapports I. Stockholm, 1960, pp. 144-164.

I. HAMMARSTRÖM. *The Price Revolution of the Sixteenth Century. Some Swedish Evidence*. In: Scandinavian Economic History Review 5 (1957), pp. 118-154.

S. HOSZOWSKI. *Europe central et révolutions des prix*. In: Annales: Economies, Sociétés, Civilisations 16 (1961), pp. 441-456.

S. IWAO. *Japanese Gold and Silver in the World History*. In: International Symposium on History of Eastern and Western Cultural Contacts (1959). Tokyo: Japanese National Commission for UNESCO.

H. KELLENBENZ. (ed.) *Precious Metals in the Age of Expansion: Papers of the XIVth International Congress of the Historical Sciences*. Stuttgart, 1981.

A. KOBATA. *The Production and Uses of Gold and Silver in Sixteenth- and Seventeenth-Century Japan*. In: Economic History Review 18 (1965), pp. 245-266.

P. LINDERT. *English Population, Wages and Prices 1541-1913*. In: Journal of Interdisciplinary History 15-1 (1985), pp. 609-634.

M. MALOWIST. *Poland, Russia and Western Trade in the 15th and 16th Centuries*. In: Past and Present 13 (1958), pp. 26–41.

W.E. MASON. *The Neoclassical Dichotomization of Economic Theory*. In: Economic Inquiry XII (1974), pp. 567–576.

– – –. *Winners and Losers: Some Paradoxes in Monetary History Resolved and Some Lessons Unlearned*. In: History of Political Economy 9 (1977), pp. 476–489.

D.N. MCCLOSKEY. *Review of Ramsey*. In: Journal of Political Economy 80 (1972), pp. 1332–1335.

J.U. NEF. *Silver Production in Central Europe*. In: Journal of Political Economy 49 (1941), pp. 575–591.

R.B. OUTWAITE. *Inflation in Tudor and Early Stuart England*. London, 1969.

G. PARKER. *The Emergence of Modern Finance in Europe, 1500–1730*. In: C.M. CIPOLLA (ed.). The Fontana Economic History of Europe. II: The Sixteenth and Seventeenth Centuries. Glasgow, 1974, pp. 527–622.

D. PATINKIN. *Money, Interest, and Prices. An Integration of Monetary and Value Theory*. 2nd ed. New York, 1965.

P. RAMSEY. *The Price Revolution in Sixteenth-Century England*. London, 1971.

J.C. RILEY AND J.J. MCCUSKER. *Money Supply, Economic Growth, and the Quantity Theory of Money: France, 1650–1788*. In: Explorations in Economic History 20 (1983), pp. 274–293.

W.C. ROBINSON. *Money, Population and Economic Change in Late Medieval Europe*. In: Economic History Review 12 (1959), pp. 63–82.

D.E. ROPER. *Two Ingredients of Monetarism in an International Setting*. In: Intermountain Economic Review 6 (1975), pp. 13–30.

F. SIMIAND. *Recherches anciennes et nouvelles sur la mouvement général des prix du XVIe au XIXe siècle*. Paris, 1932.

F.C. SPOONER. *The Economy of Europe 1559–1609*. In: R.B. WERNHAM (ed.). The New Cambridge Modern History. III: The Counter-Reformation and Price Revolution. Cambridge, 1968, pp. 14–43.

A. SZASDI. *Preliminary Estimate of Gold and Silver «Production in America 1501–1610.* In: H. KELLENBENZ (ed.). Precious Metals in the Age of Expansion: Papers of the XIVth International Congress of the Historical Sciences. Stuttgart, 1981, pp. 151–224.

A. TAKAYAMA. *Mathematical Economics.* Hinsdale, 1974.

P. VILAR. *Oro y moneda en la historia 1450–1920.* 3rd. ed. Barcelona, 1974.

V. ZIMYANI. *A Typology of Central European Inflation in the XVIth and XVIIth Centuries.* In: Journal of European Economic History (1975), pp. 399–402.

MEANING AND FUNCTIONS OF NEW METHODS OF TIME SERIES ANALYSIS FOR ECONOMIC HISTORY

by Winfried STIER

1. Introduction

Methods and procedures of time series analysis may be very important in investigating historical phenomena. For instance, the search for 'long swings' is a famous and well-known subject which has kept historians busy for more than fifty years, sometimes involving extensive use of time series-procedures. Although the purpose of my lecture is not the discussion of the question of existence (or non-existence) of 'long swings', I would like to start with a problem of time series analysis which is also important in the context of 'long swings'. It gives us a good illustration of why time series analysis is necessary and what are the main differences between modern and traditional approaches in this field. For the sake of clarity I shall first use a simulated time series. Suppose we generate a time series in the following manner. The series is additively composed of three components:

$$T_t = 0.001\ t^3 - 0.1\ t^2 + 0.5\ t + 100$$
$$C_t = 200 \cdot \sin 0.06\ \pi\ it$$
$$R_t \sim N(0,1)$$
$$t = 1, 2, \ldots, 200$$

that is:

$$X_t = T_t + C_t + R_t;\ t = 1, \ldots, 200.$$

Let us further suppose that the time index t means years, so that our time series comprises 200 years, a time span not unusual for historical time series. Of course, the first component is the so-called long-term or secular trend which gives us the direction of the general development of the series in the long range. The second component (C_t) is a cyclical one which evolves with exact regularity with a period of about 33 years. Finally, the third component (R_t) is of a different nature than the other ones. It is a so-called random-component, meaning that this component does not follow any regular pattern whatever. Its behaviour is unpredictable. This characterization implies that R_t contains no useful information for the historian. Therefore, it is often called the 'noise'-component. We assume for R_t a normal distribution with zero expectation and unit variance for all time-points. Figure 1 shows the series X_t and the components T_t and C_t.

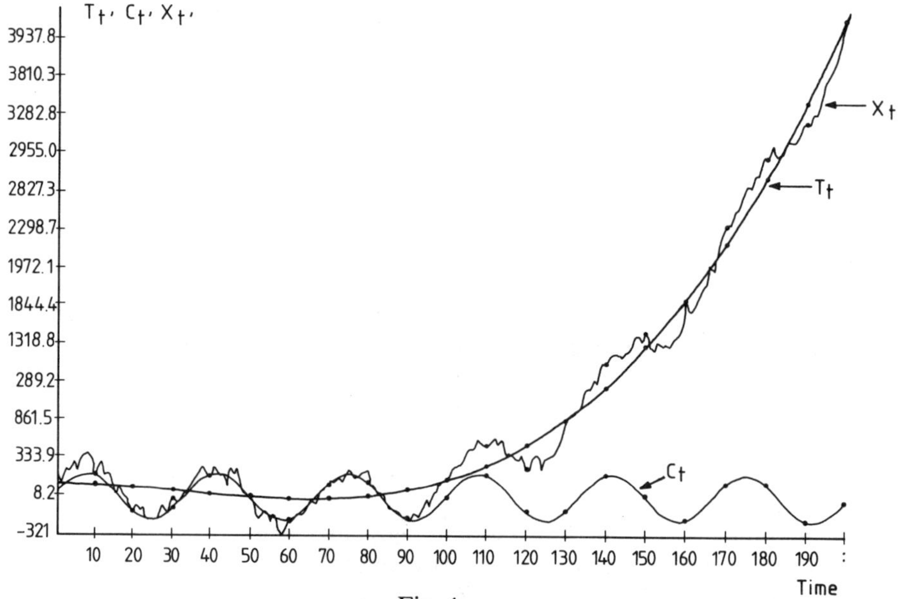

Fig. 1

Now, let us forget that we know the components of the time series and let us consider X_t to be a real time series. If X_t were a real series, we might be interested in the following questions:

- Does X_t contain a cyclical component and if yes, how regularly does it evolve in time?
- How does the trend develop in time?

Obviously, the first question seems to be the more difficult one: we have to extract a signal – namely the cyclical component C_t – which is contaminated by a second signal and by 'noise'. Of course, in our simple example, the trend is the more disturbing fact. This might be different in a different series. Now, how could we find out if the series X_t contains a cyclical component at all?

The first and easiest method would be a simple visual inspection of the given series. In our example it is not hard to see that the series does indeed contain a cyclical component whose pattern is obviously very regular at least until t = 90. After that time point the pattern seems to loose its regularity (the maxima and minima are shifted) and the amplitude of the cyclical component seems to diminish. Thus we might conclude that there is a cyclical component which, however, is not stable over the whole series and which disappears towards the end of the series. Of course, we know that this conclusion would be wrong.

It is obvious that the trend makes the visual method an insufficient one. We see also that the randomcomponent is not a severe problem here, since the variance of that component is always the same for all time-points, so the cyclical component is disturbed to the same degree for all time points. Of course, this may be quite different in real time series.

Our simple example shows that we have to eliminate the trend-component in any case before we can hope to identify a cyclical component. This example gives us a good opportunity to discuss both some traditional and some newer approaches in time series analysis.

2. Traditional Approaches in Time Series Analysis

What are the procedures usually and traditionally used in trend eliminating?

Let us begin with the so-called 'method of least-squares'. The basic idea is the assumption that the trend follows a simple function of time such as

$$T_t = a + bt$$

or

$$T_t = a + bt + ct^2,$$

or in general:

$$T_t = a + bt + ct^2 + dt^3 + \dots + mt^n,$$

whose coefficients are unknown. According to the 'least-squares'-principle, they are estimated in such a way that the differences between the values of the time series and the trend-values-squared and summed over all time-points - take a minimum. In the simplest case of a linear trend, therefore, minimize the expression

$$\sum_t [X_t - (\hat{a} + \hat{b}t)]^2 = \min_{\hat{a},\hat{b}}.$$

with respect to $\hat{a}$ and $\hat{b}$, where the 'hats' denote least-squares estimates. This approach is also called a 'regression'. After having estimated the trend-coefficients, a subtraction of the estimated trend from the given series results in a detrended series.

This brief non-technical outline of least-squares shows us the crucial points of this method. Obviously, the most delicate point is the choice of the 'correct' trend-function. What degree is to be chosen? Is a linear function sufficient or should a higher degree be chosen? It is clear that an incorrect choice of the functional form - a misspecification - has severe consequences on the resulting detrended series. For instance, if we would choose a quadratic function instead of a cubic one, the detrended series would never be similar to a regular sinusoid with constant amplitude. Unfortunately, in general we have no solid arguments which would allow us to prefer a special functional form for the trend in real time series. Neither economic theory nor history will give us reliable hints. So our choice will be more or less arbitrary. This also holds, of course, for subsequent results which depend, in turn, on the detrended series.

It shall be mentioned here that there exist special methods in time series analysis which may be useful in determining the 'adequate' degree of a trend-polynomial. However, they depend on certain assumptions which may not be fulfilled in reality. Maybe the most familiar is the so-called 'Variate-Difference-Method'. It depends on a simple fact:

Let $\Delta f(t)$ be a polynomial of degree n. Then the difference

$$\Delta f(t) := f(t) - f(t-1)$$

is a polynomial of degree not higher than n–1. For instance, let

$$f(t) = a + bt + ct^2, \quad n = 2$$

then

$$\begin{aligned}\Delta f(t) &= a + bt + ct^2 - [a + b(t-1) + c(t-1)^2] \\ &= 2ct + b - c\end{aligned}$$

and

$$\begin{aligned}\Delta^2 f(t) &:= \Delta f(t) - \Delta f(t-1) \\ &= 2c = \text{const.},\end{aligned}$$

that is, the n-th difference is a constant. This gives us a practical procedure for determining the degree of a polynomial: by differencing a series successively we finally get a series of constants. The number of differences we need until we get constants is the degree of the trend-polynomial. If the series contains a noise-component we will not get exactly a series of constants, but a series which oscillates around zero. There exists significance-tests which are useful to decide if the differenced series can be considered indeed as a pure noise-process. Details are not important here.

However, this method depends on the assumption that a series does not contain a cyclical component. If a cyclical component is present, we can difference as often as we like, but we will never get a constant or a random-series. For instance, let us consider the cyclical component in our simulated series after differencing once:

$$\begin{aligned}\Delta C_t &= C_t - C_{t-1}\\ &= 200 \sin 0.06\pi t - 200 \sin 0.06\pi(t - 1)\\ &= 200(1 - \cos 0.06\pi) \sin 0.06\pi t\\ &\quad + 200 \sin 0.06\pi \cdot \cos 0.06\pi t.\end{aligned}$$

If we perform more differences, the differenced cyclical component becomes a very complicated trigonometric expression which is not equal to a constant. So in our simulated series, the third difference is not a constant (or oscillating randomly around zero, if we include the noise-component) although the trend is in fact a polynomial of the third degree. Thus, this method is only of restricted value. Further problems of the variate-difference-method cannot be discussed here.

Let us consider some additional details and problems of least-squares. An obviously crucial point has to do with the validity of the basic time series model that is, the additivity of the components. A subtraction of the estimated trend makes sense only when the postulated additivity holds. But the components might just as well be combined in a multiplicative manner. Or a mixed model might be adequate. Since there are no dead-sure methods to discriminate between the different approaches, a further source of possible specification-error has to be taken into account.

There are other problems with the least-squaresapproach, which cannot be discussed here.

The second approach used in time series analysis for estimating a trend are the so-called 'moving-averages'. A moving-average is defined by:

$$Y_t := \frac{1}{2q+1} \sum_{s=-q}^{+q} X_{t-s},\ t = q+1, \ldots, N - q$$

where $2q+1$ denotes the length of the average and N the length of the time series. In general, it is preferable to choose an average of odd length, because the averaged values can be assigned exactly to specific time points. With even length, they would have to be assigned between time points.

The above formula shows that q values are lost at both ends of the series according to this averaging procedure. For example, if the filter length is 7, then we would loose 3 values at each end of the series. Figure 2 shows a series where the trend is estimated first by a 7-term and then by a 13-term moving average: Obviously, the longer the moving-average, the smoother the resulting trend. Since moving-averages have smoothing-properties, they are often called 'smoothers'. One problem with moving-averages involves the question: how many terms should be included? As our example shows, this is indeed a crucial problem; the smoothness of the resulting trend depends on this decision and, of course, the pattern of the detrended series also. This problem

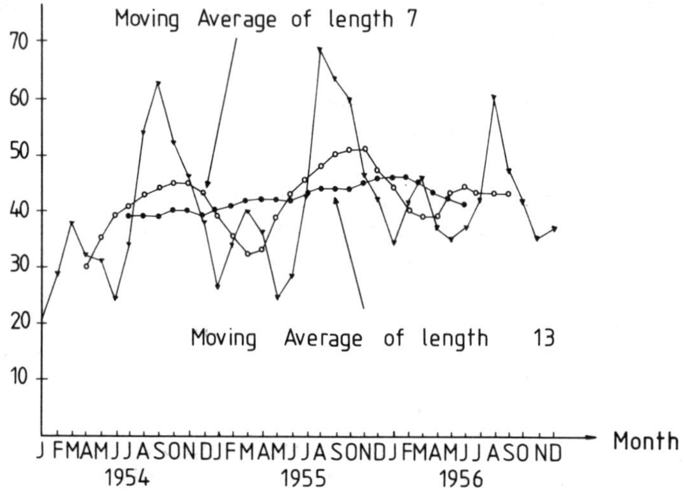

Fig. 2

is as important as the choice of the degree of the polynomial in the least-squares-approach. However, I would like to defer the discussion of this point until additional aspects of time series analysis are exposed.

Moving-averages, as introduced above, can be considered as special filters. A linear filter can simply be described as a transformation of a time series according to

$$Y_t := \sum_{s=-q}^{r} a_s X_{t-s}, \; t = s + 1, \ldots, N - r$$

with

$X_t :=$ filter-input (given time series)
$Y_t :=$ filter-output (filtered series)
$a_s :=$ filter-weights

Obviously, in the case of our moving-averages we have

$$a_s = \frac{1}{2q+1} = \text{const.}$$

$$r = q,$$

that is, they are filters with constant coefficients and they are symmetrical. This last property implies that there is no phase-shift between filter-input and filter-output. In a non-symmetrical filter or moving-average there is a phase-shift, which means that the filter-output is delayed against the given time series. This might be an undesirable feature of a filter.

Filters are important tools of modern time series analysis. Before we enter into a discussion of these concepts, I would like to make some final comments on the traditional approaches. Moving averages seem to be more flexible than least-squares. The assumption that a trend, for example, evolves according to one and the same function for all time-points seems to be a very rigid one, especially when the series is very long. Fortunately, least-squares-approaches can be modified in a way which makes them as flexible as moving averages. The modification is based on the assumption that the polynomial chosen does not hold for the complete series, but only for relatively small time-intervals. These intervals are overlapping and moving over the series. Therefore, this modification is called 'moving regression'. For example, let us assume that a polynomial of the second degree is to be fitted to the first five values of a series by least squares. Then the coefficients of the polynomial are determined by minimizing the expression:

$$\sum_t \left[X_t - (\hat{a} + \hat{b}t + \hat{c}t^2) \right]^2.$$

As solution we get for the first trend-value at $t = 3$

$$Y_3 = \frac{1}{35}(-3X_1 + 12X_2 + 17X_3 + 12X_4 - 3X_5),$$

or in general:

$$Y_t = \frac{1}{35}(-3X_{t-2} + 12X_{t-1} + 17X_t + 12X_{t+1} - 3X_{t+2}).$$

The filter-weights are (–3/35, 12/35, 17/35, 12/35, -3/35).

Note that they sum to unity. The first case, with equal weights, results by fitting a linear function.
Although least-squares can be made more flexible by using moving regressions, the basic problem, namely the choice of the functional form, remains. Additionally, there is the problem of the length of the moving regression. This is the price to be paid for greater flexibility.

3. New Approaches in Time Series Analysis

The modern approaches to be discussed here are methods which are based on filter-theory. We have learned already that moving-averages can also be considered as filters, although more elementary ones. Modern filter-theory can be characterized by the fact that no ad-hoc-solutions (like simple moving averages) are accepted, but filters are designed in such a way that they accomplish prespecified criteria. These criteria are chosen according to the

problems to be solved or the aims of the analysis.
A characteristic feature of modern filter-theory is the fact that filter-design can be made in the frequency-domain and not only in the time domain. This possibility offers a lot of advantages which will become evident when we consider details and examples.

The transition from the time domain to the frequency domain is done by mathematical operations which are known as Fourier transforms. However, I do not want to enter into this relatively complicated matter. Rather, I prefer to explain the essentials in a non-technical way. Thinking in terms of frequencies seems complicated and unfamiliar. It is easy to understand, however, if we look at well-known concepts in a little bit different way than we might be used to.

Let us begin with a monthly economic time series. Very often economic series show a seasonal pattern; for example, the unemployment series is seasonal. The seasonal component of a series can be roughly characterized by its 'periodic' nature: it repeats every twelve months. Of course, the seasonal pattern is almost never of an exact regularity in reality, since the amplitude of the seasonal component is usually changing. Therefore, we should characterize a seasonal component as a 'quasi-periodic' movement with a period of twelve months.

Now, instead of saying: 'the period of the seasonal component is 12 months' we might just as well say: 'in a time unit (which is a month here) 1/12 of the seasonal component has passed'. Or shorter: 'the frequency of the seasonal component is 1/12'. Next, let us consider the trend of a series. The trend can be characterized by the fact that it is a component which never repeats itself. That is, its periodicity is infinite or its frequency is zero. Frequencies which are between zero and 1/12 are called 'low frequencies' and frequencies beyond 1/12 are called 'high frequencies' (in monthly series).

Visually, low-frequency components correspond to smooth and slow movements in the time domain, whereas high-frequency components correspond to rough and fast movements in the time domain.

Instead of considering only two frequencies we might ask if a series contains maybe some more frequencies. Now, it may be proven mathematically that there are in fact an infinity of frequencies. More exactly, it can be shown that a series in general can be considered as being composed of a continuum of frequencies. The lowest frequency is zero and the highest is 0.5 (with discrete series).

If we design a filter in the frequency domain, we have the possibility of eliminating or preserving frequencies according to the aims of our analysis. For instance, if a series is to be detrended, the filter has to eliminate the zero-frequency. All other frequencies have to be preserved. A filter of this kind is called a high-pass filter. If the trend is to be estimated, all low frequencies below the seasonal frequency should be left unaltered, whereas the high frequencies should be eliminated. Such a filter is called a low-pass filter.

The behaviour of a filter in the frequency-domain can be described completely by a function which is usually called a 'transfer function'. This function gives us two pieces of information: the first one shows, if and how certain frequencies or frequency-bands are attenuated (resp. eliminated) or amplified. The second one shows if there is a phase-shift between filter-input and filter-output. The transfer function of a filter is generally a complex function. Its absolute value gives us the first information. Therefore, this function is called the 'gain-function' of a filter.[1]

Now, let us return to our problem of detrending a time series. As mentioned before, a high-pass filter is needed for the solution of this problem. The gain function of such a filter has the shape shown in Figure 3.

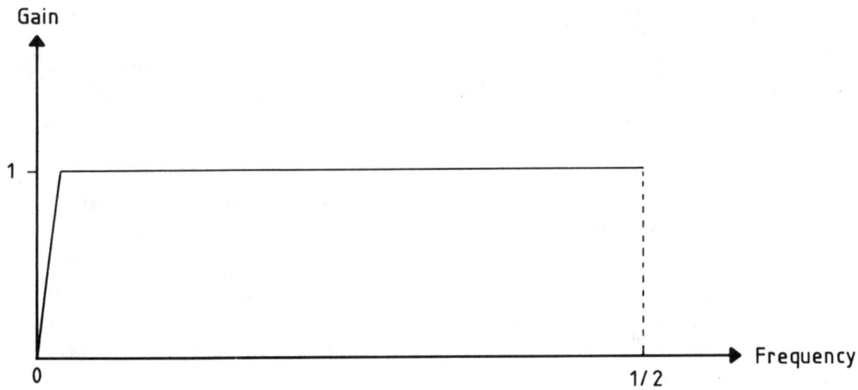

Fig. 3

A filter with such a gain function virtually eliminates the zero-frequency only and leaves all other frequencies untouched.
If this gain function is to be realized, how can we find the corresponding filter-equation? The sollution to this problem is indeed difficult. It is not possible to discuss it here in detail; only results can be given. It can be shown that realization of the gain function requires a set of three filter equations[2]:

[1] W. STIER, *Verfahren zur Analyse saisonaler Schwankungen in ökonomischen Zeitreihen*, Springer, Berlin-Heidelberg-New York, 1980.
[2] H. SCHULTE, *Statische-methodische Untersuchungen zum Problem langer Wellen*, Athenäum, 1981.

$$Y_t^{(1)} = X_t - 2\,X_{t-1} + X_{t-2} + 1.980098\,Y_{t-1}^{(1)} - 0.980487\,Y_{t-2}^{(1)}$$

$$Y_t^{(2)} = Y_t^{(1)} - 2\,Y_{t-1}^{(1)} + Y_{t-2}^{(1)} + 1.960530\,X_{t-1}^{(2)} - 0.960919\,Y_{t-2}^{(2)}$$

$$Y_t^{(3)} = 0.961057 Y_t^{(2)} - 1.922114\,Y_{t-1}^{(2)} + 0.961057 Y_{t-2}^{(2)} + 1.980098\,Y_{t-1}^{(3)} - 0.980487\,Y_{t-2}^{(3)}$$

The filter-output of the first filter ($Y_t^{(1)}$) is the filter input of the second filter and its output ($Y_t^{(2)}$) is input into the third filter ($Y_t^{(3)}$), which is finally the detrended series. The next diagram illustrates the complete filtering-process:

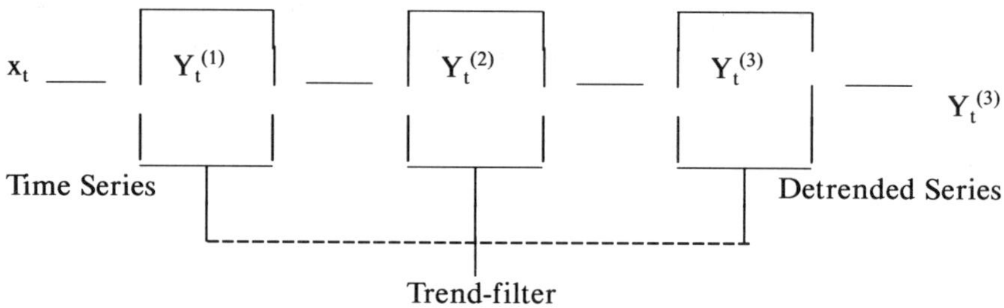

To demonstrate the efficiency of this filter, let us first consider a simulated series which is additively composed of the two components:

$$T_t = 0.001t^3 - 0.1t^2 + 0.5t + 100$$

$$C_t = 100\,(\sin\frac{2\pi}{60}\,t + \sin\frac{2\pi}{40}\,t + \sin\frac{2\pi}{20}\,t),$$

that is

$$X_t = T_t + C_t$$

The cyclical component is a superposition of three sinusoids with the periods 60, 40 and 20 time units (or with the frequencies 1/60, 1/40 and 1/20). For the sake of simplicity, the series does not contain a random component. Figure 4 shows the series and the result of detrending.

Obviously, C_t and the detrended series differ only marginally. Note, that no assumption has to be made about the functional form of the trend whatsoever.

Now, let us consider some real series. The series are English series relating to the production of consumption goods (Fig. 5), investment goods (Fig. 6), and the production of tin (Fig. 7).[3] Fortunately, the filter-design approach discussed above does not exhaust all possibilities of modern filter design. For

[3] W. Hoffmann, *Wachtstum und Wachtstumsformen der englischen Industriewirtschaft von 1700 bis zur Gegenwart*, Jena, 1940.

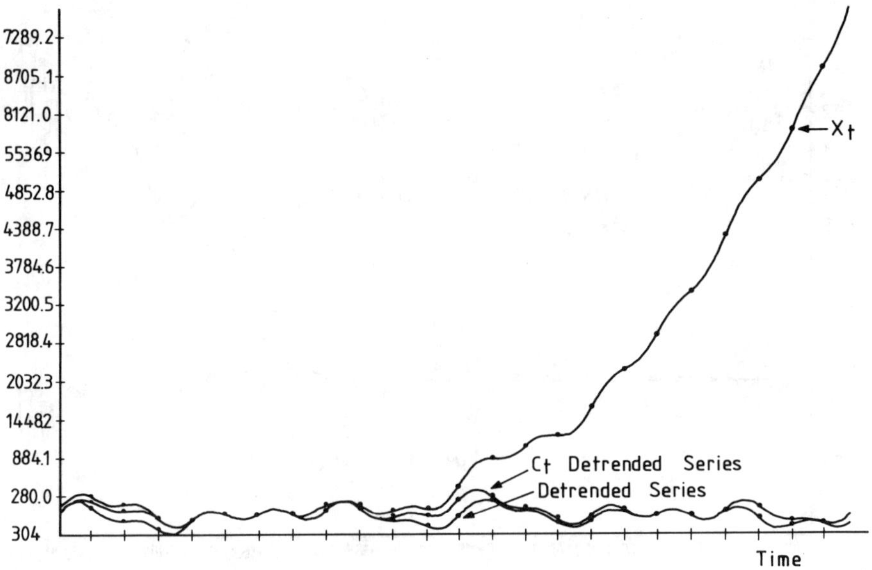
7289.2
8705.1
8121.0
5536.9
4852.8
4388.7
3784.6
3200.5
2818.4
2032.3
1448.2
884.1
280.0
304
X_t
C_t Detrended Series
Detrended Series
Time

Fig. 4

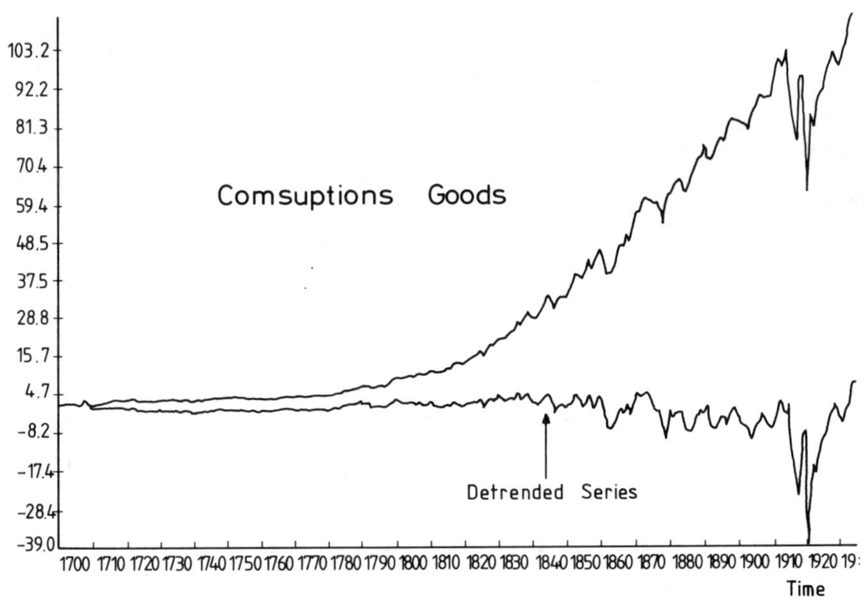
103.2
92.2
81.3
70.4
59.4
48.5
37.5
28.8
15.7
4.7
-8.2
-17.4
-28.4
-39.0
Comsuptions Goods
Detrended Series
1700 1710 1720 1730 1740 1750 1760 1770 1780 1790 1800 1810 1820 1830 1840 1850 1860 1870 1880 1890 1900 1910 1920
Time

Fig. 5

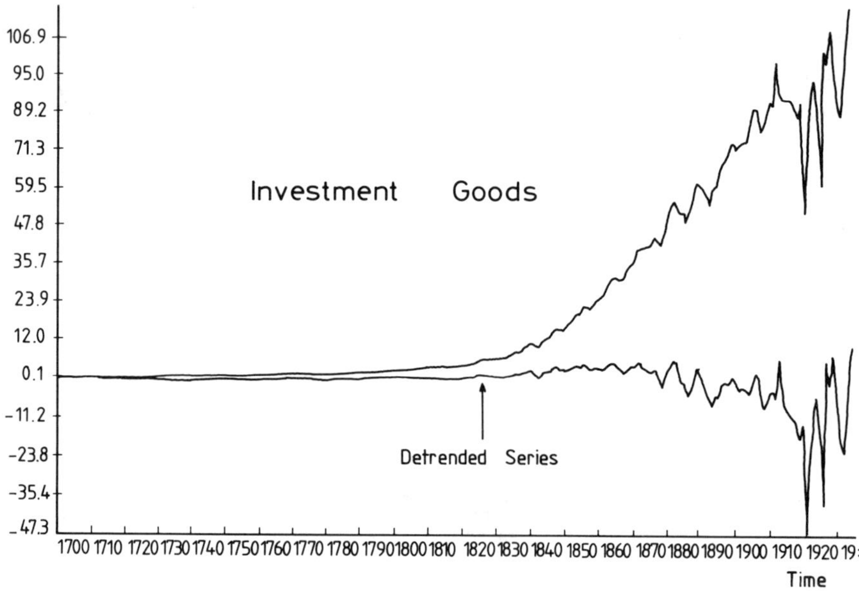
106.9
95.0
89.2
71.3
59.5
47.8
35.7
23.9
12.0
0.1
-11.2
-23.8
-35.4
-47.3
Investment Goods
Detrended Series
1700 1710 1720 1730 1740 1750 1760 1770 1780 1790 1800 1810 1820 1830 1840 1850 1860 1870 1880 1890 1900 1910 1920 19:
Time

Fig. 6

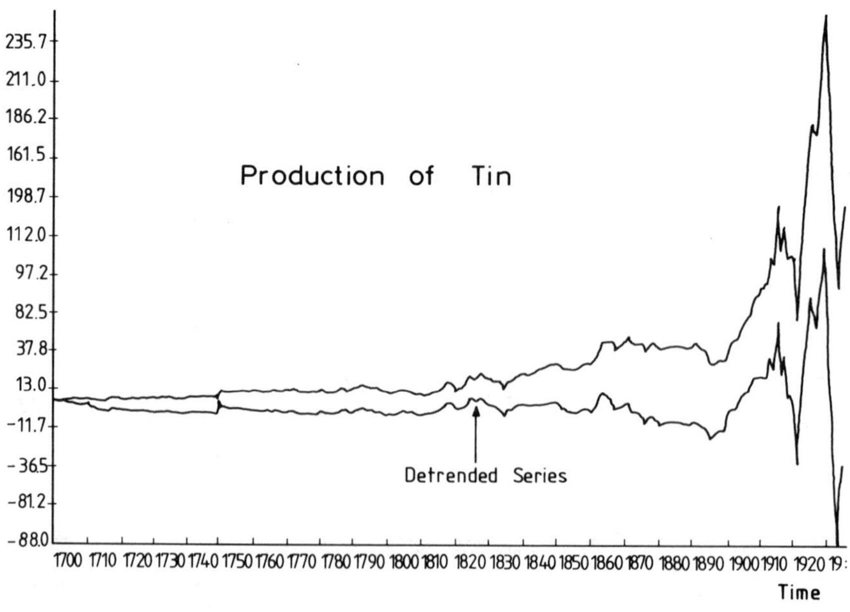
235.7
211.0
186.2
161.5
198.7
112.0
97.2
82.5
37.8
13.0
-11.7
-36.5
-81.2
-88.0
Production of Tin
Detrended Series
1700 1710 1720 1730 1740 1750 1760 1770 1780 1790 1800 1810 1820 1830 1840 1850 1860 1870 1880 1890 1900 1910 1920 19:
Time

Fig. 7

certain types of filters like low-pass or band-pass-filters (which are very important in practice), such a design will have the drawback of a remarkable phase-shift which may be disturbing in analyzing certain problems.

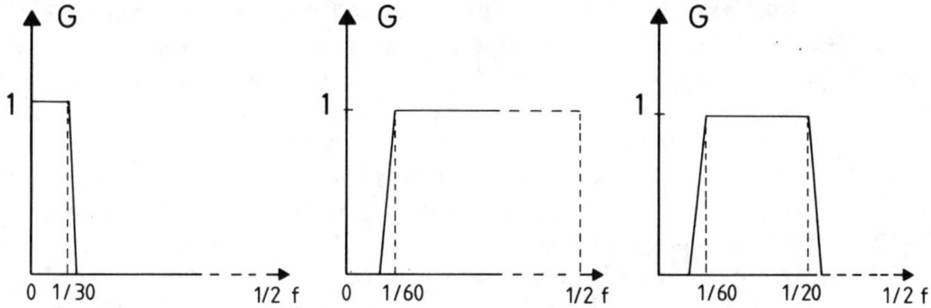

Fig. 8

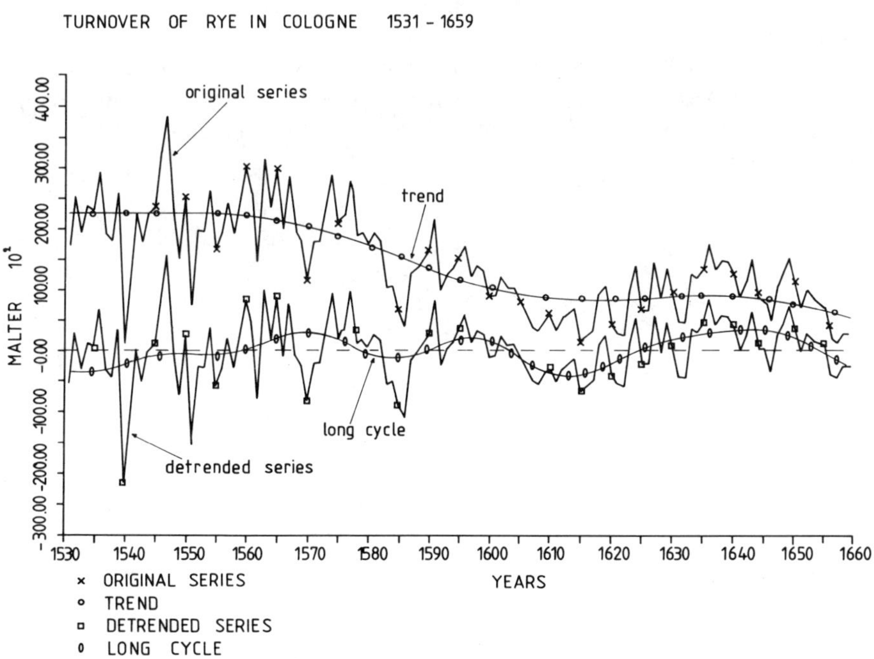

Fig. 9

This phase-shift can be avoided if we use a different approach. Its essential feature is the fact that filtering is done in the frequency-domain and not in the time-domain as in the first approach. This can be achieved by a transformation of the data into the frequency-domain by using a so-called 'Discrete-Fourier-Transform' (DFT). I shall not discuss details here. Rather, let us consider a final example. For illustration we use the series 'Turnover of Rye in Cologne (1531–1659)'. For estimating the trend a low-pass filter was chosen which preserves all oscillations with a period longer than 90 years. To estimate a trend-free series, a high-pass filter was used which is complementary to the low-pass. Finally, for analyzing any possible long cycles, a band-pass filter was taken which preserves all cycles between 20 and 60 years. The gain functions of these filters are shown in Fig. 8. The series and the results of the filtering are shown in Fig. 9.

Obviously, there is no phase-shift between the original and the detrended series. The peaks and troughs of both series are located at exactly the same time points.

These results offer the possibility to the historian of detailed investigations. For instance, he might be interested in looking at the 'long cycle'-component, its maxima resp. minima, its turning-points, the magnitude of its amplitude-values etc. Is it possible to explain them historically? Do they agree with known historical events or phenomena? If not how can the discrepancies be explained?

An analysis of these and similar questions is even more interesting and challenging when a set of series is analyzed in the way described above giving a more complete picture of historical phenomena than a single series only.